GIOTTO TO THE COMETS

Europe won its independence in space exploration with the Giotto mission. It has been a triumph of scientific imagination, technical skill, international partnership, and the human will to overcome inevitable difficulties. In bringing all these elements together in his story, in a judicious yet lively way, Calder tells what a scientific mission in space is really like.

Roger Bonnet, *Director of Science, European Space Agency*

There can be little doubt that the dual mission of Giotto has been a triumph. The encounter with Halley's Comet was 99 per cent successful; for the first time we were treated to a close-up view of a cometary nucleus, and in the space of a few minutes we learned more than we had been able to do throughout the history of science. But sending Giotto on to a different comet was more of a problem still. Remember, the mission had not been intended for anything of the sort when it was first planned.

Well, the encounter with Grigg-Skjellerup has taken place. Most of the equipment on Giotto was still working. There has been a great deal of interest, and there has been a need for a chronicler. Nigel Calder has risen to the occasion. His book is lively, readable and informative. He takes us through the entire Giotto story, with all its hitches, its delays and its final success; he writes in a way which will interest both the general reader, with no prior knowledge, and the specialist who needs a reference book which can be consulted quickly and easily.

In short a most interesting book about a most interesting and important mission.

Patrick Moore

Nigel Calder's best yet! A gripping – and true – adventure story about a twelve-year mission into the unknown. Here is the high drama, the politics, the emotion, the incredible characters, that lie behind the smooth facade of space exploration. But in the end there is just one hero – the brave, battered and inspirational Giotto space probe.

Heather Couper

GIOTTO TO THE COMETS

NIGEL CALDER

PRESSWORK

1992

Published by Presswork
47 Gartmoor Gardens, London SW19 6NX

First published 1992

ISBN 0 9520115 0 6

Printed in Great Britain by Bath Press.

CONTENTS

AUTHOR'S NOTE

Where wast thou when I laid the foundations of the Earth?

The purposes of Giotto's voyages to Halley's Comet in 1986 and Comet Grigg-Skjellerup in 1992 lay deep in fundamental science. As a result a worldly tale of men and women engaged in a high-tech adventure turns out to encompass cosmic mysteries, implied in God's question to Job. The European Space Agency offered me the chance to tell the story of its spacecraft Giotto, however I liked. It granted me free access to individuals and documents. I am grateful to the managers, scientists, engineers, operations staff and flight dynamicists within the agency who spared time to reminisce about the mission and update me on its progress and technicalities. The opinions and interpretations are my own.

Thanks are due to many members of the Giotto science working team in France, Germany, Ireland, Switzerland and the United Kingdom, who welcomed me to their laboratories and explained their theories, experiments and results. Individuals from British Aerospace and Dornier patiently educated me in spacecraft engineering. Alan Johnstone of the Mullard Space Science Laboratory and Rod Jenkins of British Aerospace assisted me with their personal collections of papers on Giotto and Halley's Comet, and David Dale with his personal photo album. Johannes Geiss (Bern), David Hughes (Sheffield), Jean-Loup Bertaux (Paris) and Tobias Owen (Honolulu) were also generous with advice and information. For both of Giotto's cometary encounters, I was present at the European Space Operations Centre, Darmstadt.

Teamwork is the essence of the mission and the following remark, in Chapter 3, is worth anticipating as an apology to protagonists and a reassurance to the reader:

'If the author named all of the hundreds of individuals and dozens of organizations involved in the science, engineering and operations of Giotto, and who deserve a mention, this book would read like a telephone directory. The reader is asked to take it for granted that everyone named was part of a larger team.'

Nigel Calder

ABBREVIATIONS AND COSTS

Space literature is typically spattered with initials and acronyms. These have been kept to the minimum that is compatible with common sense. NASA, more familiar than the names of some of the world's smaller nations, is used without gloss or apology for the US National Aeronautics and Space Administration, and corporations best known by their initials (RCA, BBC, etc.) are left in that format.

Other abbreviations which avoid long-winded repetitions within passages, or have a special meaning in context, have also been adopted. The following occur quite often. The geographical references in the first three cases often serve as convenient alternatives:

ESTEC: European Space Research and Technology Centre, Noordwijk.
ESOC: European Space Operations Centre, Darmstadt.
JPL: Jet Propulsion Laboratory (NASA), Pasadena.
DSN: Deep Space Network (NASA).
CCD: charge-coupled device.
zulu: Universal Time or Greenwich Mean Time.

Figures are quoted for the costs of projects only when they have some direct importance in the story. Inflation and variable exchange rates make intercomparisons, and conversions to national equivalents, tricky or misleading. The European Space Agency uses 'accounting units'. The conversion rates to US dollars in two relevant years were:

1986: 1 ESA accounting unit = $ 0.73
1992: 1 ESA accounting unit = $ 1.15

The cost to the European Space Agency of sending Giotto to Halley's Comet in 1986 was 144.4 MAU or $ 104 million. The total budget of the Giotto Extended Mission to Comet Grigg-Skjellerup in 1992, including the first reactivation and Earth flyby, was about 12 MAU or $ 14 million.

These figures do not include the costs of the scientific experiments. Whereas in NASA projects the agency typically pays for those, in European projects the investigators are funded by their institutes and national science organizations. There is no global figure for the experimental costs of Giotto.

ILLUSTRATIONS

Photographs not otherwise credited are from the European Space Agency

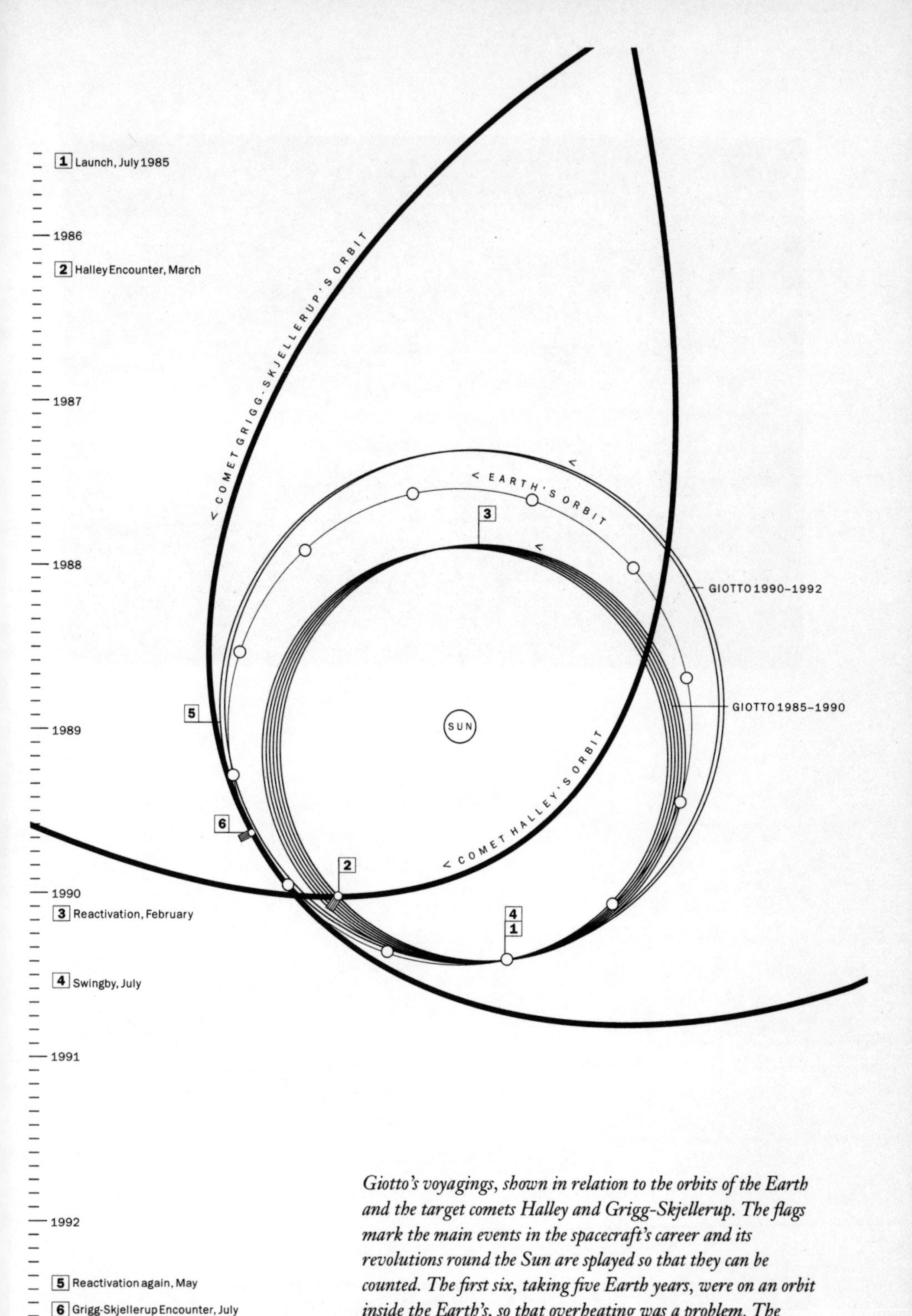

Giotto's voyagings, shown in relation to the orbits of the Earth and the target comets Halley and Grigg-Skjellerup. The flags mark the main events in the spacecraft's career and its revolutions round the Sun are splayed so that they can be counted. The first six, taking five Earth years, were on an orbit inside the Earth's, so that overheating was a problem. The Earth swingby in 1990 put Giotto in a wider, chillier orbit.

Halley's Comet, photographed on 9 January 1986 with the 80-centimetre Schmidt telescope of the Calar Alto Observatory, Spain. The part of the tail visible in this photograph extended for about 6 million kilometres. (Observation by Kurt Birkle, Max-Planck-Institut für Astronomie, Heidelberg.)

CHAPTER ONE

THE LURE OF HALLEY'S COMET

'I WONDER IF YOUNG UWE HERE WOULD LIKE TO SEE A COMET.'

The astronomer Karl Wurm was a friend of the Kellers. Though he and they lived in bomb-ruined Hamburg, they were all Bohemian Germans who fled from their homes at the end of bloodstained Europe's worst of wars. When the Red Army stormed in from the east to meet the allies from the west, Uwe Keller was a bewildered and hungry toddler. Twelve years later, in the spring of 1957, the round-faced schoolboy was keener on engineering than astronomy, but a chance to look at the sky from the Hamburger-Sternwarte, up the hill at Bergedorf, was not to be scorned, and Professor Wurm seemed unusually excited.

With his eye at the telescope, Uwe saw at once that a comet did not rush across the sky, as many people imagined. A pointed smear of light was moving in a stately fashion in front of the stars. In fact you could see stars through it. While Wurm took photographs of the comet, called Arend-Roland, he explained that it was a cosmic dust cloud far flimsier than cigar smoke but brightly lit by the Sun. There was gas in the comet too.

The pressure of sunlight on the powdery grains pushed them away from the comet, creating the tail that stretched for millions of kilometres through space, until it became too thin to see. Some of the grains would show up later if they collided with the Earth atmosphere, streaking as meteors through the upper air. A comet's tail always pointed away from the Sun, and Arend-Roland was travelling tail-first because it was leaving. But its 'anomalous tail' was causing a stir among comet watchers.

'Look, Uwe, at that protrusion from the head! It is rare sight, but exceptionally visible in this comet. You are looking edge-on at a fan of large dust grains, on the sunward side of the comet's tail.'

Wurm invited the young Keller to return the next night, and the next. The boy saw emphatic changes in the comet's position. He learned that comets came in from the depths of space, swung around the Sun, and then flew off. Arend-Roland would not be seen again, Wurm said, but some comets settled into elongated orbits around the Sun that brought them back at intervals of a few years or a few centuries. Out of five to ten comets spotted each year, about half were old friends and half were newcomers.

'When Halley's Comet came in 1910, astronomers looked for a lump, a nucleus, in the middle of the coma, the comet's head. But they could see nothing like that. So they said, "Well, the comet is just a swarm of corpuscles, like grains of sand, that travel together around the Sun." But many of us now believe that there really is a nucleus. We don't see it because it may be only a few kilometres wide, and millions of kilometres away.

'The latest theory from America,' Wurm added, 'says that a comet is a ball of ice contaminated with dust. Then, when it comes near the Sun, some of the ice vaporizes, carrying dust with it and making the enormous display you can see out there.'

That was when Uwe Keller first observed a comet. If real life were like a storybook, he ought to have set his heart, there and then, on discovering the dirty snowball in the heart of a comet. Instead, Keller forgot about comets for ten years and returned to them only by an elongated mental orbit.

While Comet Arend-Roland headed for the outer darkness, the Soviet Union put the first manmade satellite, Sputnik-1, into orbit around the Earth in October 1957. The US followed suit four months later. Apart from the important military uses of satellites, there was prestige to be won, as the superpowers competed in a circus of ever more amazing exploits.

Scientific goals made the space race more respectable, and Soviet and American researchers seized the chance to send instruments aloft. They discovered the Earth's radiation belt and looked at the Moon's far side. Astronomical telescopes went into orbit to detect ultraviolet and other rays hidden from view on the ground by the Earth's atmosphere.

Among the European scientists who craved a piece of the action was a mop-haired French student with a large smile, Roger Bonnet. He was so flabbergasted by Sputnik-1 that he resolved to make a career in space exploration. He attached himself to Jacques Blamont in the Service d'Aéronomie at Verrières-le-Buisson, in the improbable setting of a 19th-Century fortress in a forest on the outskirts of Paris. Blamont proposed instruments for American spacecraft, and planned experiments with French-built rockets and satellites. When France launched its first satellite in 1965, it became the third independently spacefaring nation.

Other nations thumbed a ride, and NASA hoisted European satellites into orbit, beginning in 1962 with a British spacecraft and continuing with satellites created by the fledgling European Space Research Organization. But the Americans were patronizing, and likened Europe's hesitant efforts in space to 'small boys talking about sex'. While the enthusiasm of Bonnet and a few others found an outlet, for most young European scientists spaceflight remained a spectator sport.

Uwe Keller had given up the idea of engineering. At Hamburg University, he was seduced by physics; not the fashionable theme of subatomic particles but the astrophysics of the Universe at large. Keller went on to Munich, as a graduate student, sporting a beard and a slower car than he would have liked. He started work on the atmospheres of stars.

The smart place to be, if it would have you, was out on Munich's Föhringer Ring where Werner Heisenberg, discoverer of the uncertainty principle of atomic physics, had his own laboratory: the Max-Planck-Institut für Physik und Astrophysik. Its top man in astrophysics was Ludwig Biermann, Germany's best-known comet scientist, who had predicted the wind from the Sun.

While a professor at Göttingen, Biermann had been puzzled by the gassy tails of comets. The electrified gas, or plasma, glowed blue and was not very obvious to the human eye. Poring over astronomers' photographs of plasma tails, Biermann was struck by how neat and straight they were, compared with the dust tails which often sprawled and curved.

What guided the luminous plasma tails? The pressure of sunlight, which drove the dust grains, would not affect atoms or molecules in a gas sufficiently. Biermann visualized a breeze of electrified particles, another plasma, blowing out from the Sun and through the Solar System. It would sweep the comet's plasma in a direction away from the Sun. Physicists already blamed magnetic storms and radio blackouts on the Earth on puffs of particles from the Sun. The continual streaming of the plasma tails of comets required, Biermann said, a non-stop solar wind.

This prediction was one of three theories, all promulgated in 1950-51, that modernized comet science. Fred Whipple, at Harvard College Observatory in the US, developed another of them, in suggesting that the comet nucleus was a confection of ice and dust. Comets often returned a little early or late, and Whipple reasoned that emissions of dust and gas from the dirty snowball, acting like a jet engine, could alter a comet's orbit slightly.

At Leiden in the Netherlands, Jan Oort asked himself, 'Where do comets come from?' He calculated that there were billions of comets slowly orbiting in a cloud far from the Sun. When a passing star perturbed them, it would send some falling into the inner Solar System, where people could see them. The Whipple snowball, the Oort cloud and the Biermann wind gave research on comets a new framework of ideas.

The solar wind was the first to be confirmed, in an early scientific triumph of the Space Age. As soon as probes with suitable instruments left the Earth's neighbourhood to explore the interplanetary void, they found that 'empty space' was filled with the Sun's hot breath. When astronauts went to the Moon, they set out traps for particles of the solar wind.

Biermann was a quiet, absent-minded man, a scientist through and through. He always turned the conversation back to physics and would make calculations on the back of an envelope, relying on logarithms which he knew by heart. When asked to sum up his accomplishments he would recite the names of his students. In 1967 Biermann took on his last student: Uwe Keller.

He set Keller a task in comet science. If Whipple was right, and the nucleus of a comet consisted largely of ice, then most of the gas coming from it would be water vapour. The Sun's rays would break up the water, and Biermann predicted that a large cloud of hydrogen atoms should form around a comet's head. He suggested that Keller develop this theory in more detail.

No one had seen a hydrogen cloud accompanying a comet. The gas would emit ultraviolet rays that could not reach telescopes on the ground. Early in 1970 the American satellite OAO-2, an orbiting astronomical observatory, detected the hydrogen cloud around Comet Tago-Sato-Kosaka. The French space scientist Jacques Blamont, and a young colleague Jean-Loup Bertaux, had an ultraviolet sensor in another American satellite, OGO-5. They turned it on the bright Comet Bennett and again registered the hydrogen.

As Keller's work on Biermann's idea provided a theory for interpreting the space observations, Blamont and Bertaux visited him in Munich. They showed him the comet data and found the young German's comments shrewd. Blamont invited Keller to the fortress laboratory at Verrières-le-Buisson, where he helped Bertaux to work over the observations. The hydrogen cloud surrounding Comet Bennett was enormous – 15 million kilometres wide.

Keller saw that the future of comet studies lay in space. When Munich declared that he could call himself Doctor Keller, he swapped the Alps for the Rocky Mountains. NASA's generous spending on science was luring many Europeans to the US. At the University of Colorado, Keller worked on space instruments to scrutinize comets from sounding rockets and satellites. He joined US Navy scientists in equipping the manned spaceship Skylab for comet observations, and so discovered the hydrogen cloud of Comet Kohoutek at the end of 1973. On his return to Germany, Keller became involved with plans for a US-European astronomical satellite, International Ultraviolet Explorer.

In 1976 he quit Munich and drove north again, to join the Max-Planck-Institut für Aeronomie. This was at Katlenburg-Lindau, a village of timber-framed houses in the Leine valley between the university city of Göttingen and the Harz mountains. Just up the road at Duderstadt the tanks and nuclear artillery of NATO and the Warsaw Pact faced one another across a geopolitical fault-line. A side road led to the institute's modern building, stocked with computers and space-age workshops but standing amid woods and meadows like a medieval monastery. Here Keller began to think about Halley's Comet.

In 1684 Edmond Halley, a Londoner, visited Isaac Newton in Cambridge. He found that Newton had a theory that described how the Sun's gravity governed the planets. It crowned the efforts of Copernicus the Pole, Kepler the German, and Galileo the Italian, to create a modern science of astronomy. Halley had to chivvy Newton into publishing the theory of gravity, and the very bright comet of 1680, which both men had observed, became a test case.

In medieval cosmology the Earth was the centre of the Universe and the Sun, the Moon, the planets and the stars revolved around the sky every day, with angels presiding over the celestial clockwork. Comets were supposedly close to the Earth and, as unruly apparitions, the work of the Devil. In 1577 Tycho the Dane compared sightings of a bright comet from different parts of Europe and proved that it was much more distant than the Moon. But Kepler caused many years of confusion by asserting that comets travelled in straight lines.

Newton saw that the 1680 comet veered sharply as it passed the Sun. Its orbit carried the comet out to an incalculable distance, yet in the Sun's vicinity it plainly obeyed the law of gravity, like all the better behaved objects in the Solar System. In the new machinery of the heavens, comets made sense.

'I am out in my judgment,' Newton declared, 'if they are not planets of a sort, revolving in orbits that return into themselves with a continual motion.'

This was Halley's cue to look into the records of past comets to see whether any showed up more than once on the same orbit. After much figuring he offered three repeaters. Two were incorrect, but he was surest about the third. In 1682 Halley had observed a comet going around the Sun the 'wrong' way, opposite to the Earth's revolution. It matched apparitions recorded in 1607 and 1531, and possibly a comet of 1456, less well-documented.

'Whence I would venture confidently to predict its return, namely in the year 1758,' Halley declared.

By the appointed date, Halley was in his grave but his forecast was not forgotten. As the months of 1758 slipped by, the watchers were reassured by French calculations indicating that it might be a little later than Halley said. On Christmas Day a German amateur, Johann Palitzsch, saw the smudge in the sky and Europe cheered a triumph of the new astronomy. Halley's Comet had come back.

It returned in 1835 and 1910, and was due again in 1986. Between the apparitions, as scientists came to realize, mementos of the comet took the form of meteor showers. Every October and early May, the planet passed near the orbit of Halley. The litter of dust grains that it left behind hit the Earth's atmosphere and burned up as shooting stars.

In the late 1970s, all comet scientists were looking forward to Halley's next visit. The curator of its orbit was Donald Yeomans of NASA's Jet Propulsion Laboratory in Pasadena. Helped by a colleague in Ireland, Yeomans made computations and searches through the records to establish an unbroken sequence of twenty-nine apparitions of the Comet Halley, from 240 BC to 1910. The intervals varied from 76.1 to 79.3 years, because the gravity of the Earth and other planets perturbed the comet's orbit. The retrospective identifications showed, for instance, Chinese astronomers logging it in 240 BC and the Babylonians in 164 BC.

In classical and medieval times, when astronomers usually doubled as astrologers, comets foreshadowed famine, plague, war and the death of kings. After Halley appeared large and close to the Earth in AD 837 the Chinese emperor classified astronomy as secret. In 1066 Halley was taken as a portent of the Norman conquest of England, and was embroidered on the Bayeux Tapestry which chronicled the events. As Edmond Halley suspected, the comet of 1456 was the same object.

'Its head was round and as large as the eye of an ox,' said Leonardo da Vinci's teacher, Paolo Toscanelli. 'Its tail was prodigious for it trailed through a third of the firmament.'

In 1910 the progress of science heightened public alarm about Halley, rather than allaying it. Astronomers had identified the poisonous gas cyanogen in comets' tails, and the Earth was to pass right through the tail of the comet on 19 May 1910. Despite assurances that any poison would be imperceptible, quacks managed to sell pills as antidotes.

The writer Mark Twain was born and died under successive apparitions, in 1835 and 1910, but most people could expect to see Halley just once in a lifetime. Periodicity, superstition and science combined to make it special. As the apparition of 1986 drew near, the chief cataloguer of the comets, Brian Marsden, assured astronomers that the public was animated by three objects in the Solar System: the planet Mars, the rings of Saturn and Halley's Comet.

For scientists, Halley's importance lay in its predictability. Great new comets sometimes surpassed it in their vigour, and their untarnished condition made them ideal objects for study, but any observations were necessarily hasty. Of all the recurrent visitors, Halley was by far the most active, and preparations for watching it could be made years in advance. Predictions for the 1986 apparition showed that it would be the poorest for many centuries to the naked eye, but modern telescopes and spacecraft could mount an unprecedented campaign of observation. The public would see the comet well enough, on television.

Comet science was at a low ebb and badly needed the new knowledge and new funding that only Halley's Comet could deliver. Electric street lighting made all comets unimpressive to city-dwellers. Only a lunatic fringe still thought they were telegrams from Hell. And the general view among professional astronomers in the 1970s was that comets were garbage.

Discoveries in a frame far grander than the Solar System preoccupied them: galaxies rent by explosions, pulsating neutron stars that foreshadowed the existence of black holes, and microwave radio energy pervading space as a relic of the Big Bang itself. If comets were a swindle, as the snowball theory indicated, and generated their spectacular displays from a tiny nucleus, they were a mere pollution of the sky.

Worldwide, the professionals who regarded comets as their main interest would barely fill a bus. Amateur astronomers were the real enthusiasts, willing to spend their nights searching for the speck that might grow into a great comet and bring fame to its discoverer. But once a newfound comet's orbit was computed, and any peculiarities noted, there were few original observations worth making from the distance of the Earth.

Even the planets were too parochial for most professional astronomers. When the Space Age made possible the first on-the-spot inspections of the Moon and the other planets, much of the planning came from geophysicists rather than astronomers. They saw a chance of understanding the Earth better by comparing it with its companions. Investigators of meteorites, the stony and metallic lumps that reached the ground from outer space, were another group active in lunar and interplanetary missions. But to suppose that the nebulous comets might have at least as much to say about the Earth's history as pieces of Moon rock, or close-up pictures of the surface of Mars, defied preconceptions. As for a space mission to a comet, had not Whipple himself been urging it in vain since 1959?

The devaluation of comets was perverse and untimely. If the nucleus of a comet was pitifully small, its gravity was very weak. Its contents would therefore avoid the fierce modification by heat and pressure experienced by all the stuff of the Earth and other planets. So the comets offered a direct link to the wider cosmos, and to the once-upon-a-time when the Sun and the Earth came into being.

A hint came with fluffy grains from outer space, when a NASA research aircraft collected them from the stratosphere. Typically a hundredth of a millimetre wide, the alien objects looked like fish-roe. They were pieces of some comet's tail, and small and light enough to drift slowly through the atmosphere instead of burning up as meteors. Their ingredients resembled the stony parts of carbon-rich meteorites, and their loose, fluffy structure spoke of weightless realms in the deserts of space, where physical and chemical events were far gentler than on Earth.

Just how well could science account for the creation of the Universe and its contents, right down to human beings? That was always an apt test of the state of knowledge. In the 1970s, the scientists' version of the first chapter of Genesis was persuasive in some parts, but shaky in others. Comets could help to fill the gaps.

Interstellar dust preserved in comets might specify how the chemical elements were made, in stars that existed before the Sun and the Earth came into being. The comets themselves were raw materials left over from from the building of the planets, and a rain of comets may have helped to stock the planet's atmosphere. In one hypothesis, the Earth was built in its present region of the Solar System but its ocean arrived from somewhere beyond Uranus, in the cometary mail.

Whether such ideas were daft or brilliant, who could tell? The total obscurity of the origin of life on the Earth left room for bold conjecture, and two astronomers in the United Kingdom, Fred Hoyle and Chandra Wickramasinghe, asserted that comets were alive with bacteria.

'Our argument,' they wrote, 'is that life arrived eventually on the Earth by being showered as already living cells from comet-type bodies.' They explained life's evolution by reinfections with comet dust. Few experts took them seriously, although their reasoning about 'prebiotic' chemicals in interstellar dust, deliverable to the Earth by comets, won more attention.

Another notion often considered wild was that a collision between the Earth and a comet wiped out the dinosaurs. Yet in the late 1970s, in gorges near Gubbio in Italy, European geologists found a layer of red clay where most living things in a former ocean suddenly expired, just when the dinosaurs ashore were dropping dead. Californian scientists analysed the clay and found strange abundances of the element iridium, a rarity at the Earth's surface but commoner in meteorites. Opinions were divided as to whether the impactor was a comet or a small planet, an asteroid, which could itself be an ex-comet.

Doomsday for the dinosaurs was only one of the mass extinctions that kept changing the course of animal evolution, and space exploration confirmed that cosmic mayhem was probably to blame. Most solid bodies in the Solar System turned out to be pockmarked by impact craters, except those like the Earth, where the crust repaired itself. Even here, satellite images helped to increase the count of impact craters still discernible on the Earth to over a hundred. Their ages sometimes matched abrupt changes in the fossils, from one stratum to the next.

Discoveries of many comets and asteroids jaywalking in the inner Solar System, each large enough to cause a major disaster, reinforced these inferences from past events. Some scientists suggested that to protect the Earth against their impacts would be a healthier use for nuclear weapons than any others contemplated.

Faced with colourful interventions by specialists from other fields, comet scientists were not so much antagonistic as aloof. They wished for a more orderly approach to their subject. Proposals about the origin of the ocean, or of life, or of deadly impactors, depended on conceptions of the comet nucleus for which the only evidence was indirect.

Yes, the dirty snowball was popular, but eminent predecessors of Fred Whipple had for much longer supposed that a comet was 'a bagful of nothing' – a swarm of grains following similar orbits around the Sun. And even a snowball nucleus could just as well melt away entirely, as become an Earth-threatening asteroid. The theory-spinning seemed premature, to put it politely, at least until a spacecraft could confirm the existence of the nucleus and discover its chemical qualities.

Scientists advising NASA wanted to inspect Halley's Comet at close quarters during its visit to the Sun in 1986. But rival ideas for interplanetary voyages competed for limited funds, and to win authorization for a mission could take as much intellectual and emotional energy as executing it. American space scientists set themselves high standards of excellence. The Viking missions to Mars were a resounding success in 1976, and in 1977 an epic tour of the outer planets by the Voyager spacecraft was about to start.

For those interested in comets a fleeting flyby was not good enough. What you ought to do was make a rendezvous with the comet, by putting your spacecraft into the same orbit and letting them fly in company for several months. The true rendezvous became a fixed idea among American comet scientists, but in the Halley case it would require huge amounts of rocket energy, because of the comet's habit of going around the Sun in the opposite direction to the Earth.

All spacecraft departing for deep space inherited the Earth's orbital speed in addition to anything contributed by their rockets. Usually it was an advantage, but a spacecraft that was to fall into step with Halley had first to cancel this earthly motion and then build up a high speed the other way. No ordinary chemical propellant would do, but NASA's engineers had another idea: an electric rocket that would gradually accelerate a spacecraft for years on end, while drawing its power from large panels of solar cells. They hoped that the lure of Halley would win federal funding for this solar-electric propulsion system. But the last possible launch date was 1982 and by January 1978, when no funding was in place, even the most gung-ho engineers had to admit that time had run out. Scratch the rendezvous with Halley's Comet.

The comet scientists came up with another scheme that would postpone the launch until July 1985. It would use solar-electric propulsion to effect a true rendezvous with Comet Tempel 2 in 1988. On the way, it would fly at high speed across Halley's bows. The spacecraft would be too delicate to risk in a close encounter. It could take pictures, though, and launch a small instrumented probe to fly through the comet's dusty head.

With their minds set on the leisurely rendezvous with Tempel 2, and being scornful of a high-speed encounter, the Americans offered the dirty part of the mission to the Europeans. NASA invited the European Space Agency to provide the Halley probe that would ride piggyback on the main American spacecraft heading for the Tempel 2 rendezvous. A European team was expected in Washington on 11 October 1978, to review the scheme with their NASA counterparts. Were Europe's space scientists ready to join a mission to Halley's Comet? Yes, officially they were – though only just.

In July 1978 a US launcher lifted into space a European satellite called Geos-2 which was to investigate the behaviour of subatomic particles within the Earth's magnetic shield – the magnetosphere. A problem with the launcher had put its predecessor, Geos-1, into the wrong orbit, so the launch of Geos-2 was a tense affair. An hour after lift-off, success seemed assured.

British Aerospace was the prime contractor for the satellite and David Link, the project manager, took advantage of this euphoric moment at mission control to sow a thought in the mind of Ernst Trendelenburg, the European Space Agency's director of science. Spare parts existed for building Geos-3. Why not use it, Link said, for further explorations of the magnetosphere, and then send it onwards to a comet?

This was not the first proposal of its kind. The Germans had built two Helios spacecraft, launched by the Americans in the mid-1970s for a deep space mission close to the Sun. The German government floated the idea of adapting Helios for a joint European mission to Comet Encke. But space agencies preferred to find spacecraft for missions, rather than missions for spacecraft.

The European Space Agency had ten member states in 1978: Belgium, Denmark, France, Germany, Italy, the Netherlands, Spain, Sweden, Switzerland and the United Kingdom. Ireland was interested in becoming a full member. Scientists from any of these countries could put up ideas for missions, and a queue of proposals was always awaiting attention. Each represented the passionate desires of one group or another, and the selection process often ended in rage or grief. Anyone enthusiastic about a comet mission had to convince fellow-scientists who knew little about comets and cared less.

The fight for a European mission to Halley's Comet started in 1977 in the agency's Solar System working group, where active researchers from various countries had to agree on which projects to recommend for preliminary study. The chairman was Johannes Geiss from Bern in Switzerland, who strongly advocated a comet mission. He knew NASA's plans, and European participation in a Halley flyby was already mooted informally. Also on the table was an independent comet project using Europe's new Ariane launcher, then under development in France.

Geiss therefore asked another friend of comet research, Hugo Fechtig of Heidelberg in Germany, to work up proposals about Halley to put to the working group. Geiss and Fechtig shared the scientific background needed for making the case for a comet mission. Geiss came to international prominence as a space scientist when US astronauts deployed his 'Swiss flag' on the Moon – a metallic foil that collected heavy atoms in the solar wind. When returned to Bern, the contaminated foil yielded traces of various elements to mass spectrometers, which weighed and counted individual atoms. The Swiss physicists use the same analytical technique on samples of Moon rock, as well as on meteorites.

At the Max-Planck-Institut für Kernphysik in Heidelberg, Hugo Fechtig's cosmophysics group likewise analysed meteoritic and lunar material by mass spectrometry. Geiss and Fechtig also examined interplanetary dust found in the Earth's upper atmosphere, which came from comets' tails. While rival groups of scientists argued for space missions to this planet or that, Geiss and Fechtig's broad view of the Solar System enabled them to explain the special value of a Halley mission. More clearly than some of their counterparts in the US, these European physicists saw comets as unrivalled samples of the Solar System's primordial material, which promised clues to our cosmic origins.

The comet itself made the material accessible for analysis by spewing bountiful gas and dust into space. And although a fast flyby would leave little time for observing it, Geiss and Fechtig knew there was a beautiful bonus. The mass spectrometers, which would sort the atoms and molecules, required them to be fast-moving, but the relative speed of spacecraft would provide swift particles without any artificial acceleration. And high-speed dust grains, spattering into an instrument, would vaporize spontaneously into molecular fragments suitable for chemical analysis.

In September 1978, after meetings with other European scientists, Fechtig formally proposed that the Solar System working group should recommend a study of a Halley flyby, which could be with the Americans or without them. By then, Geiss was no longer chairman. The group could choose just one project for detailed study, and the comet lost out to the Moon. British scientists wanted a satellite in orbit around the Moon, to map its chemistry, and the project virtually guaranteed good scientific results. But a high-speed flight through a comet lasting only a few hours? In the US, even the comet specialists scorned it! A fierce debate ended with a vote. Four hands went up for Halley, six for the lunar mission.

'Gentlemen, the argument is over,' the chairman said. He was wrong.

An ardent comet scientist, Jean-Loup Bertaux, rounded on one of those who had voted with the majority – a British specialist on planetary atmospheres.

'Why this sudden interest in the Moon?' Bertaux demanded. 'It has no atmosphere!'

'It's all politics, don't you see?' came the embarrassed reply. The scientist had felt obliged to support his fellow-countrymen.

Bertaux turned to Fechtig and they discussed tactics. They could not stop the group's verdict going to the science advisory committee, which was the next battleground, but Bertaux could tackle the committee's chairman, Roger Bonnet, a colleague from the Verrières-le-Buisson forest. Fechtig would lobby a comet enthusiast on the committee, Ian Axford from the Max-Planck-Institut für Aeronomie in Germany. They would put the same simple argument: the Moon was always there, but the next chance to go to Halley would not occur until 2061.

The science advisory committee met a few days later in a hotel at Nice airport in France. Although its members were respected scientists, the committee had a reputation in academic circles for merely endorsing decisions of the European Space Agency's officials. As a new chairman, Bonnet was determined to change that. He therefore asked the agency's director general and director of science to leave the room while the committee examined afresh the issue of Halley versus the Moon.

The officials were happy to wait in the bar. The director general wanted robust science advice which would improve the agency's standing in the universities of Europe. And although the rules required the director of science, Trendelenburg, to pass on the Solar System working group's choice of the lunar mission, he hoped to see it reversed.

His was the hidden hand that mattered. Known for his bald head, flashy ties and a well-stocked drinks cabinet, Trendelenburg was quick to decide if individuals were competent, whether they were bosses above him or raw young scientists joining his team. To those who passed his test he gave unstinted loyalty and trust; anyone else had a hard time. He had seen more difficulties than promise in a comet mission, until he paid a visit to his ageing mother in Germany.

'So Ernst, what are you doing these days?' she asked.

Trendelenburg spoke of various projects that he and his teams were studying, but they meant little to her. She was no scientist.

'Some crazy people even want to inspect Halley's Comet,' he added.

'Halley's Comet – that's out of the ordinary! I saw it as a child.'

'That was in 1910.'

'Yes, but I remember it as if it were yesterday – an astounding sight. And now you have the chance to send a craft to look at Halley's Comet from nearby!'

Trendelenburg confessed privately to his friends that his mother's reaction altered his perspective. In their closed, highly technical world, scientists could forget the wide-eyed interest of non-scientists in the wonders of nature. When he sensed how popular the comet mission would be, he decided that the technical snags would simply have to be overcome.

As he waited in the bar in Nice that day, Trendelenburg wondered if he would be able to accept NASA's urgent invitation to examine the proposed Halley flyby and Tempel 2 rendezvous. Under Bonnet's leadership the science advisory committee duly showed its independence of officialdom. It deferred the lunar mission and chose the comet proposal for further study. Paradoxically it thereby did just what one official, Trendelenburg, wanted.

Supporters of the lunar mission were outraged. Before they could organize a protest, Trendelenburg used his mandate from the committee to embark on intensive joint studies with the Americans, for the design of a European Halley probe to ride on NASA's mother ship, in what came to be called the International Comet Mission. He shelved Fechtig's alternative of an independent mission, and mustered a team to go to Washington on behalf of the European Space Agency for talks with NASA about Halley.

The North Sea, grey in autumn, beat on the dunes that protected the tulip fields of Noordwijk from the salt water. A few kilometres inland stood Leiden. From this Dutch city of art and science the Pilgrim Fathers set off for an American wilderness, just when Rembrandt was dropping out of the university to learn to paint. In modern Leiden the astronomer Jan Oort concocted his theory of a vast cloud of comets surrounding the Solar System.

The seaside dunes supported the exclusive links of the Haagse Golf Club, which Rüdeger Reinhard, a young scientist arriving from Kiel in Germany, had the wit to join when joining was still possible. They sheltered the modern buildings of the European Space Research and Technology Centre, where Reinhard worked in the space science department. ESTEC, as everyone called it, was the main laboratory of the European Space Agency.

Scientific spacecraft came into being through a marriage of engineering and science. Nuptials were routine at Noordwijk, where multinational colonies of engineers and scientists shared the same campus. Following a practice that served NASA well, scientists initiated the missions, in consultation with top experts, but engineers commanded them as project managers. The link between the management and the experimenters who would use the spacecraft was the project scientist appointed from the space science department.

In that autumn of 1978, Reinhard's thoughts were far away in the atmosphere of the Sun. ESTEC scientists were required to stay active in their own research, and Reinhard was working on the theory of energetic particles from solar flares. He raised his eyes from his desk and saw a burly, square-jawed man leaning on the doorpost of his office like a truculent cowboy in a Hollywood Western.

'Where's Reinhard?' the stranger demanded. His accent was English.

'That's me.'

'I'm Dave Dale and I'm supposed to work with you.' The deep voice rang with authority and verged on the unfriendly. Reinhard prided himself on his command of English, and found a nuance of dignified consent.

'I expect we can work together,' he said.

'Well what do you know about Halley's Comet?' Dale asked.

When Reinhard stood up, he was taller than Dale, but slimmer, with a boyish face. He had heard that the Englishman was heading a new office for studying future projects. Of Reinhard, Dale knew only that he was a German scientist. You had to work with all kinds of people at Noordwijk, but there was no rule that said you had to smile.

David Dale had joined ESTEC four years earlier, from a British defence establishment where he worked on military satellite communications. He was assigned to a small team studying an international solar-polar mission, later called Ulysses. A European spacecraft was to be launched by NASA, to swing around Jupiter and go into an orbit over the poles of the Sun.

'Where's Jupiter?' Dale asked, exposing his ignorance of the Solar System.

With common sense and shrewdness he held his own among scientists and engineers whose paper qualifications far surpassed his own. The Ulysses project took him often to NASA's Jet Propulsion Laboratory in California. He then worked for two years on the payload for Europe's Spacelab, which fitted into the cargo bay of the US space shuttle. Spacelab work kept Dale away from Noordwijk, until the European Space Agency's director of science asked him to take charge of technical studies on future projects. The proposal that the agency should provide the Halley probe for the International Comet Mission landed on Dale's desk.

The team to go to Washington to open talks with NASA would include a study scientist. ESTEC's scientific hierarchy named Reinhard. He did not let Dale's forbidding manner put him off, but threw himself into the comet project with enthusiasm. At an argumentative meeting at the Dornier aerospace company, which made an engineering study of the Halley probe, Reinhard fought gamely on the project's behalf. When they were returning from Dornier by ferry across the Bodensee, Dale apologized to Reinhard for his doubts about him. The two main champions of Europe's comet mission laughed and became firm friends.

The concept of the Halley probe evolved over twelve months, amid much transatlantic commuting. Dornier based its design on an existing spacecraft called ISEE-2. The probe would have just fifteen days of independent life, after release from the American mother ship, so its housekeeping facilities were rudimentary. Its signals had to reach only as far as the mother ship, which would relay them to the Earth. Batteries would provide power during the encounter with Halley.

The American spacecraft, with its solar-electric propulsion system taking it onwards to Comet Tempel 2, would use its motors or flywheels to preserve its orientation. The Halley probe, stabilized more simply by being spun like a rifle bullet, would enter the head of the comet with its spin axis pointing along the line of approach. A bumper shield in front would protect it from the impacts of comet dust. The probe would be aimed to pass within 500 kilometres of the supposed nucleus of the comet.

In February 1979, in the old Bavarian city of Bamberg, a US-European meeting defined the probe's scientific payload – its suite of instruments. David Dale and Rüdeger Reinhard were there of course, and Marcia Neugebauer from the Jet Propulsion Laboratory (JPL) spoke for the main American mission. Jean-Loup Bertaux and Uwe Keller were among seven scientists from European and American academic laboratories.

The Europeans gave top priority to three instruments for analysing the chemistry of the comet's gas and dust. Other devices would study the interaction of the comet with the solar wind and count the impacts of dust grains on the bumper shield. No camera figured on the list at first, because the European side thought that it would be too heavy, and obtaining good images from the spinning probe without smearing them would be very tricky. The mother ship would take pictures from a safe distance.

It was a persuasive astronomer, Mike Belton from Tucson, Arizona, who convinced his colleagues from Europe that the probe should have a camera of its own. It should see the comet's nucleus at least well enough to tell the experimenters where the probe was, from moment to moment, and so help them to interpret their results. It might also show the size and shape of the nucleus. A novel concept for an electronic camera offered by Alan Delamere of the Ball Aerospace company in Boulder, Colorado, seemed to solve the spin problem.

Adding the camera worried Keller. He understood the technical challenge of obtaining images, but saw little scientific challenge in interpreting them. Didn't a picture give you its information almost at a glance? There was more 'real science', he thought, in analysing a complex string of data from a particle counter or a mass spectrometer.

He had worked on the chemical composition of comets for ten years, and his heart was in that instrumentation. He was afraid that the camera, under American leadership, would take too much of the probe's payload. To prevent that happening – in effect, to cramp its style – he enlisted in the camera team and obtained funding so that his laboratory could contribute to the work. Keller did not know what he was letting himself in for, by this backhanded engagement with a space camera.

Reinhard, as the study scientist, collated the proposals for instruments and also found himself designing the probe's bumper shield. The kiss of Halley's Comet could be fatal. Dust from the comet, hitting the probe at a relative speed of fifty-seven kilometres per second, would sandblast it severely, and a large grain could destroy it. Dale and the engineers wanted precise estimates of the dust hazard.

JPL had a rough dust model based on observations of Comet Encke. An inter-agency working group developed it, with the European Space Agency paying for computations at JPL that predicted the numbers of dust grains of different sizes, at different distances from the comet's nucleus. Experimenters in the US shot small grains at aluminium sheets, at several kilometres per second, to assess the damage that small meteorites might do to spacecraft. The Ernst-Mach-Institut at Freiburg in Germany performed similar tests for the European Space Agency.

Fred Whipple had invented a dual bumper shield to protect a spacecraft against small meteorites. A fairly thin front sheet stopped the smallest grains, while the impact exploded any grains large enough to penetrate it. The vaporized material then spread out in a gap, before arriving at a second sheet which should be able to absorb the blunted impact.

Reinhard learned the principle of the dual shield at JPL. He figured out the practical requirements in time for a workshop on the dust hazard at Noordwijk in April 1979. His proposal was for a front sheet one millimetre thick, separated by a gap of twenty-five centimetres from a rear sheet to be made of aluminium honeycomb as thick as the weight constraints would permit. Reinhard's calculations became the basis of the shield's design.

Fears that impacts would surround the probe with a glowing sheath of electrified particles also troubled the scientists and engineers. High-speed molecules and dust grains could set up electric charges and voltages which could damage the spacecraft or its instruments, and confuse the scientific observations of the comet. When Reinhard coordinated studies of this problem, it began to look less alarming than some supposed.

Dale was eventually satisfied that the probe could be built to match the Americans' requirements and expectations, and the chief wishes of European scientists. Its total mass was to be 143 kilograms, with forty-eight kilograms allotted to the scientific instruments. On 23 October 1979, in a formal 'announcement of opportunity', NASA and the European Space Agency invited scientists to offer experiments to the International Comet Mission. Researchers in the US and Europe aroused their thoughts and their teams. A notice of intent had to be filed with the agencies by 23 November.

In its issue dated 26 November 1979, just three days after that initial deadline, the aerospace magazine *Aviation Week* carried a devastating news item. The US Office of Management and Budget had deleted solar-electric propulsion from NASA's funding. With the cost of the space shuttle soaring, the budget- makers had rejected all of NASA's requests for new programmes save one, a gamma-ray observatory. As *Aviation Week* also reported:

'NASA's comet science working group has told the agency that should the Halley flyby/Tempel 2 rendezvous mission be killed . . . a flyby to Halley alone would be "unacceptable".'

The chairman of that comet group was Joseph Veverka of Cornell University. In a newsletter, called *Comet Chronicle*, he drummed up support for the threatened mission. Telling how NASA's administrator met with President Jimmy Carter in December 1979, to plead in vain for the reinstatement of solar-electric propulsion, Veverka identified Carter's science adviser as the culprit to whom expressions of 'amazement and chagrin' should be directed. *Comet Chronicle* also listed congressional committee members who could vary the budget allocations. The lawmakers were unmoved.

The American comet scientists had promoted a mission more ambitious than anything contemplated in Europe, the Soviet Union or Japan. But as the Halley deadline approached, their European friends diagnosed collective folly. The price tag on the International Comet Mission was large enough to catch even the president's eye. The small band of comet scientists had neither the power to command the money nor the tact to beg for it.

Imprudently, they were trying to play politics with the politicians. The assertion that a new propulsion system was the only way to achieve a true rendezvous with a comet was technically incorrect. And the comet faction in NASA offered Halley's Comet as the bait to fish for the solar-electric propulsion that would take their spacecraft to a quite different comet. In the same breath they said that the Halley flyby was futile, and suitable only for the Europeans and other backward peoples.

Yet all other interplanetary exploration had begun with flybys. And by saying that a Halley flyby alone was 'unacceptable' the American comet scientists spurned several conventional options open to NASA. They would have only themselves to blame if they were left without any mission to Halley's Comet.

CHAPTER TWO

A MISSION FOR EUROPE

Trois, deux, un . . . feu!

On Christmas Eve 1979, in a forest clearing in French Guiana, rocket engineers lit a candle for Europe. Lemon-coloured flames scorched the launch pad and the noise congealed the tropical air. Four motors, burning dimethylhydrazine in nitrogen peroxide at a rate of a tonne of propellant per second, heaved the 210-tonne Ariane launch vehicle off the ground.

If nothing went wrong, Ariane's first launch would end a sorry tale of fecklessness in space technology. In 1964 the European Launcher Development Organisation set out to combine a British first stage, a French second stage and a German third stage in a rocket called Europa. But nobody was clearly in charge and when, by 1972, Europa had failed four times in four test launches, there was nobody to sack except the entire misshapen organization.

Out of the wreckage of Europa the European Space Agency was born. The new agency absorbed the much more successful European Space Research Organisation and took responsibility for launcher development too. The engineering of a new rocket would be done more coherently, with a prime contractor to bang the table if any sub-contractors in the member states fell down on the job. It was obvious that the French would take the lead.

Cultural diversity was one of the strengths of the new collaborative Europe that slowly pieced itself together in the decades after the Second World War. France could be relied on for a sense of glory that was highly unfashionable in other countries. The United Kingdom abandoned serious rocketry when the US sold it ballistic missiles. There, as in other parts of Europe, many politicians regarded civilian spaceflight as an extravagant circus better left to the superpowers.

The Germans, whose own experts from the wartime V2 rocket programme were kidnapped to serve in both the American and Soviet military and space establishments, were technically adventurous but politically cautious. And although a remarkable number of governments subscribed to the European space enterprise, under pressure from their scientists and their high-technology industries, most of them saw it as a cheap, cost-sharing alternative to the financial black hole of home-grown spaceflight.

The French view of space was entirely different. Denied military favours from the Americans, France made its own ballistic missiles in a crash programme in the 1960s. Its expertise in rocket engineering was unequalled in Western Europe. And the French reasoned that anyone who took a pride in technology needed a presence in space, which they saw as a potential source rather than a sink of wealth.

Satellites were becoming indispensable for telecommunications, weather forecasting and observation of the environment, and European agencies and companies were busy with their development. To the French, dependence on the superpowers for launchers was intolerable. But self-reliance was not enough. Europe should aim to beat NASA in the fast-growing commercial market for satellite launches.

'Wait and see,' the French rocketeers said. 'We'll soon be putting up American satellites.'

The French paid half the costs of preparing the Ariane launcher for the European Space Agency. Their Centre National d'Études Spatiales became the prime contractor. The Aerospatiale company of Toulouse acted as industrial architect, for the design and technical management of the project. Germany was the next largest subscriber to the Ariane programme, and other member states made contributions and pieces of the rocket. But the countdown was in French.

Ariane's first stage burned for 140 seconds, and dropped away. The engineers held their breath. The second stage ignited and burned for 135 seconds. By then the rocket was travelling at five kilometres per second, high over the Guiana coast. As the third stage, powered by liquid hydrogen and oxygen, took the speed up to ten kilometres per second after a burn of nearly ten minutes, the guidance system placed a satellite exactly into a prescribed orbit that carried it high above the Equator. At the right distance from the Earth, a small motor in the satellite injected it into an Earth-synchronous orbit, such that the spacecraft appeared to hover over one spot on the Earth's surface. It was an operation often done by the Americans, but never before by Europe.

The complete success of the first test launch of Ariane was something to celebrate over Christmas. It could not have been timelier for the scientists who were fretting about the demise of the International Comet Mission, because Ariane offered a launch vehicle for an independent European mission to Halley's Comet.

Ernst Trendelenburg, as director of science, had grown weary of American condescension. Until that moment the European Space Agency had always been the junior partner in its dealings with NASA. Now he saw a chance for Europe to do something highly visible, spectacular even, where the Americans had given up. Informal contacts in Washington confirmed the *Aviation Week* report on the demise of the International Comet Mission. Even before NASA officially informed their partners of its cancellation, Trendelenburg was considering alternatives.

He had the classic advantage of the tortoise. The NASA hare could outrace anyone, but governmental whims from year to year could break its stride. Although the European Space Agency's budget for space science was far smaller, and Trendelenburg always felt short of cash, the money trickled in from the member states continuously and predictably. He saw no deep problem about funding a comet mission that would take only a fraction of his budget over a period of six years.

Trendelenburg knew, though, that the European Space Agency faced a severe test. As an intergovernmental enterprise, its policymaking and budgeting were closely watched by the makers of policies and budgets in its member states. In this bureaucracy of bureaucracies, prudent but time-consuming procedures involved a hierarchy of committees, phased studies of possible missions, and equitable interactions with the aerospace companies of Europe. Getting a mission authorized could take several years, if it did not fall by the wayside.

Halley was already nearer than the planet Saturn and gathering speed every day for its appointment with the Sun in February 1986. The last date for a launch to intercept it was in the summer of 1985. Five years to design and prepare the spacecraft would not be generous, and was already six months less than for the International Comet Mission. To achieve a go-ahead by July 1980, Trendelenburg had to bypass cherished procedures, without antagonizing either the space science community or the governments of member states. Both groups had the power of veto.

The proposal for an Ariane-launched mission to Halley had languished while the NASA collaboration lasted, but the files contained a suggestion from an Italian space engineer, Giuseppe Colombo of Padua. Colombo was respected in the US as the inventor of crazy-seeming but effective schemes for redirecting spacecraft by swinging them around planets, or for tying satellites together with a long tether. His bald head and grey moustache were familiar in the corridors of the European Space Agency too. In March 1979 Colombo gave substance to the idea advanced by British Aerospace a few months earlier, when it pointed out that pieces for a third Geos satellite existed at its factory at Bristol.

'Send Geos-3 on a dual mission,' Colombo said. 'It can explore the Earth's magnetotail and then go on to Halley.'

By the magnetotail he meant the invisible comet-like wake of the Earth in the solar wind. One advantage of the idea was that most of the cost could be shuffled into the agency's programme on the solar wind, which commanded wider support than comets among space scientists. The Halley interception would occur in March 1986 after it had rounded the Sun, in distinction from the pre-perihelion encounter in late November 1985 expected in the International Comet Mission. Colombo offered a name for the mission: HAPPEN, for Halley Post-Perihelion Encounter.

The spacecraft would fly through Halley's tail, about a million kilometres from the nucleus. The Geos spacecraft was not well-equipped for observing the comet and Colombo suggested only that a 'meteorite detector' be added to observe the dust. An internal report supporting HAPPEN said there was no need for a camera because it would see no more than a good Earth-based telescope.

To administrators and scientists, including Trendelenburg, who knew little about comets, HAPPEN looked like a neat way of reaching Halley, as an exercise in interplanetary gymnastics. At ESTEC in Noordwijk, the Geos project scientist favoured HAPPEN. British Aerospace put in an unsolicited bid to supply the spacecraft.

Trendelenburg knew the tricks of the science administrator's trade. Careful wording of the minutes of a meeting could shade the impression given of opinions and conclusions. Cost estimates for missions were another tool for guiding policy. The margins of uncertainty prevailing in advance of serious engineering studies gave officials scope for emphasising possible cost overruns in projects they wanted to discourage. For the comet mission, Trendelenburg's tactic was to stress how cheaply it could be done.

His first move failed abjectly. When he put HAPPEN to the agency's Solar System working group at Noordwijk on 24 January 1980, it rejected it. The group's members were not against a comet mission, like their predecessors who had preferred a lunar mission just sixteen months earlier. On the contrary, they thought that HAPPEN did not do justice to scientific interest in the comet. Those who knew the plans for the International Comet Mission's Halley probe thought there was no comparison. 'HAPPEN is far less exciting,' one said.

Rüdeger Reinhard and his cometary friends wanted a mission dedicated solely to Halley, equipped especially for comet research, and sent into the head of the comet, not its tail. Jean-Loup Bertaux suggested trying for a joint flyby-only mission with NASA. But Reinhard knew Trendelenburg's mood, and thought an all-European project remained the better hope.

As an employee of the European Space Agency junior to the Geos project scientist, Reinhard was barred from any overt manoeuvre. But at the psycho -logical moment the phone rang and it was Uwe Keller calling Reinhard from Katlenburg-Lindau. He had heard the news about HAPPEN's rejection.

'What do we do now?' Keller asked.

'I advise you to send a telex to Trendelenburg right away,' said Reinhard. 'Have Geiss, Fechtig and everyone put their names to it.'

'What should it say?'

Reinhard outlined the contents for the message and dictated some key phrases. 'I can't sign it myself,' he said. 'Officially this has nothing to do with me.'

Keller drafted and polished the text. Although it was going from one German scientist to another, he wrote in English because that was the everyday language of the European Space Agency. Then he quickly gathered signatures. Hugo Fechtig of Heidelberg was happy to confirm his support for a mission he had long advocated. So was Keller's teacher, Ludwig Biermann of Munich. By the time scientists down the corridor in the Max-Planck-Institut für Aeronomie had signed it, Keller was able to append eighteen names to the telex, besides his own.

Transmitted from Keller's laboratory in the curious e.e. cummings style of a telex, without capital letters, the message arrived in Trendelenburg's office in Paris on 29 January. After noting the 'surprising loss of American cooperation' and suggesting that Halley was the only comet this century active enough to be an outstanding target for a flyby mission, the telex proposed a crucial modification to HAPPEN. Why not build two Geos-type spacecraft, one for the magnetotail and one for the comet? The same Ariane rocket could launch them both. Failing that, the Colombo scheme for a single spacecraft should be modified by putting cometary instruments into Geos-3. The message concluded:

'The first interplanetary mission of ESA could be accomplished in connection with a well-supported Earth magnetotail mission for relatively small additional costs.'

Keller's telex delighted Trendelenburg. He gave David Dale's office just one week to prepare technical remarks on a pure comet probe launched jointly with Geos-3. This was the true moment of conception for the European mission to

Halley's Comet. Reinhard was called in, to give scientific colour to work that his own advice to Keller had evoked.

Trendelenburg put the new proposal to the agency's science advisory committee in Paris on 6 February. Its chairman was now Klaus Pinkau, a German physicist. This time, the chief project competing with any comet mission was a magnificent scheme of French astronomers, in which a satellite called Hipparcos would map the positions of stars far more accurately than ever before.

The comet mission had the air of being primarily a German project. Hipparcos was strongly backed by the French government, which thought it did not have enough of a say in Europe's space science programme. When the committee's former chairman, Roger Bonnet, was invited to describe the studies of the Halley probe undertaken with NASA, he spoke so warmly of its possible all-European successor that the French delegation to the European Space Agency upbraided him for lack of loyalty.

Like the Moon, a previous competitor of Halley, the stars could wait. The comet had its own inexorable schedule. The science advisory committee made stern provisos about the quality of the instrumentation, technical feasibility and costs, but then it unanimously supported the dual mission, Geos plus a separate comet spacecraft, as the next scientific project of the agency. But Dale had to validate the mission quickly, or Hipparcos would take its place.

'The decision was made in such a short time that the opponents couldn't rally,' Keller commented. 'I have my doubts if we would have got Giotto if we had gone through the normal selection procedures.'

The mission acquired its name because an article by an American art historian in the previous year pointed out that the earliest realistic portrait of Halley's Comet was in Italy, at Padua, in a painting of the Adoration of the Magi by Giotto di Bondone. It was in a well-known cycle of frescoes in the chapel of the Scrovegni family, but no one had commented before on the identity of the comet that served as the star of Bethlehem. In the Middle Ages, when comets were bad news, this was the pleasantest role that anyone found for them.

As Halley put in one of its more sensational appearances in 1301, Giotto could not have failed to see it, and the manner of its representation was typical of the naturalism that he introduced into medieval art. Modern astronomers who examined the painted version said that it was so faithful to the naked-eye appearance of a great comet that Giotto must have observed it intently.

Rüdeger Reinhard drew the attention of others in the European Space Agency to the article, and Giuseppe Colombo, from Padua himself, suggested to a fellow-countryman in Ernst Trendelenburg's office in Paris that the European spacecraft going to Halley's Comet should be called Giotto. The name won instant acceptance. It began to crop up in official documents in February 1980, and distinguished the dedicated comet mission from previous schemes, including Colombo's own HAPPEN.

No other spacecraft ever had so magical a name. Some were cursed with mere initials, like OAO or IUE. Others had astronomical names (Kosmos, Venera) or mythological ones (Apollo, Ulysses) or invented portmanteaus (Intelsat and Landsat). A broad category of names were of anonymous travellers (Explorer, Voyager).

The magic of 'Giotto' came from its sound. By honouring the first Italian master painter in the bow shock of the Renaissance, the name showed that physicists were not uncultured folk. Yet there was nothing pompous or cute about it. 'Giotto' tripped off the tongue like a nickname, hinting at pizza and red wine. To an aluminium drum festooned with gadgets it gave a personality.

The comet mission had not yet reached the top of the greasy pole of the European Space Agency's selection procedures. The science advisory committee could only authorize the agency's staff to initiate detailed studies of a proposed mission. When those were ready, they had to go to the science programme committee, where research officials from all the member states made the decision to build a spacecraft.

The delegates were too experienced to be impressed by arm-waving enthusiasts. When the Giotto study team presented its proposals on 4 March, they thought them half-baked. Yes, the speed of response to the collapse of the International Comet Mission was laudable, but quality mattered too. The delegates registered dissatisfaction by reinstating Hipparcos as the next scientific project.

They gave the Halley project one last chance. The Hipparcos schedule might be stretched to make Giotto affordable too, and to simplify matters the science programme committee aborted Geos-3, Giotto's supposed fraternal twin. Cost-saving should come from launching the spacecraft in company with someone else's satellite. Instead of automatically using Ariane, the agency was to invite NASA to contribute the launcher, and ground links with the spacecraft, in return for a share of Giotto's scientific experiments. If the study team could tell a more convincing story within four months, the comet mission might yet fly.

'But don't bother to come back if Giotto costs more than 80 million accounting units,' the delegates said.

Dale took personal charge of the study, to the virtual exclusion of his other duties at ESTEC. He won a big concession from headquarters. As Giotto would be based on the Geos spacecraft, Dale could deal directly with its builders, British Aerospace, for an industrial input into the study. That saved him the rigmarole of putting a study contract out to competitive tender, as agency rules normally required, and it foreshadowed another time-saving decision. If the mission was approved, British Aerospace would probably get the contract for design and construction.

Dale brought a Frenchman, Robert Lainé, into his team as the chief project engineer. Work on Europe's astronomy satellites had earned Lainé a reputation for ingenuity and impatience. Dale was becoming obsessed with the mission to Halley. For a project manager, it was in every sense the chance of a lifetime.

'If we play our cards right, Robert,' he said to Lainé over a beer during that hectic summer of 1980, 'we can finish up managing Giotto ourselves.'

Dale relied on Rüdeger Reinhard for the same fanaticism on the science side. The International Comet Mission, with a similar launch date, had begun recruiting proposals for experiments half a year earlier. Although Giotto was not yet approved, a 'preliminary announcement of opportunity' went out to the scientific community in April. Reinhard also had to sharpen the presentation of the case for Giotto.

The hierarchy of scientific committees had already guided the agency away from HAPPEN, via Giotto plus Geos-3, towards the pure Giotto mission. At the end of May, the panels of scientists provided a useful intermediate examination for Reinhard and his colleagues, who had to explain the basic questions that Giotto was designed to answer. What is a comet made of? In what forms and abundances do its materials gush into space and decompose? How does the comet's atmosphere interact with the solar wind? How big is the nucleus?

Some were enchanted to learn for the first time how Giotto's very high speed through the comet, so often said to devalue the mission, promised the beautiful bonus of pre-accelerated particles for the chemical instruments. The intellectual battle was won, and the doubters swayed or silenced. The Solar System working group said it was 'unanimously enthusiastic', warm words from a panel often rent by dissensions. Two weeks later the science advisory committee strongly endorsed the recommendation that the Giotto mission should fly.

While Reinhard already had a starting list of instruments, carried over from the International Comet Mission, Dale was looking at a very different spacecraft to carry them into the heart of the comet. The old Halley probe was required to survive for only two weeks after it left the mother ship, but Giotto had to make an eight-month voyage into the depths of space. Many new questions arose, about power supplies, housekeeping, control and communication.

Engineers always had to persuade the client in advance that there were no problems, and then live with consequences as the work proceeded and the problems multiplied. Dale's team at Noordwijk and the British Aerospace engineers at Bristol had time to do little more than block out the phases of the mission, relate them to the subsystems of a spacecraft somewhat like the existing Geos, schedule the development programme to meet the launch date, and make an honest guess at the costs. An overestimate would jeopardize Giotto's chances of acceptance, while understating the costs would cause grief later.

A glaring weakness compromised the very idea of an all-European mission. When Giotto reached Halley, it would be so far away that none of Europe's ground stations would be able to receive its signals clearly. In any case, the agency's flight dynamics experts said that the spacecraft would be on the other side of the Earth at the hour of the encounter. NASA's Deep Space Network of large radio dishes included one near Canberra, but the Americans might demand too high a price, financially or politically, for using it.

Dale went to Sydney to bargain with the Commonwealth Scientific and Industrial Research Organization, whose radio astronomers had a 64-metre dish at Parkes in the outback of New South Wales. The European Space Agency would give Parkes improved microwave equipment, Dale said, if it could be the downlink station for Giotto at its crucial phases. The Australians said they'd love to be part of such a great project, and it would be a fillip to engineers who sometimes grew weary of sitting out there in the bush, swatting flies while the astronomers listened for quasars.

When Dale arrived back in Europe, just before the fateful meeting with the science programme committee, Trendelenburg greeted him with sombre news. NASA was asking 10 million dollars for the use of the Deep Space Network in the Giotto mission. Dale laughed and told Trendelenburg that Parkes was available for an expenditure of 200,000 dollars. They had a fine coup to report to the delegates.

The dream-factory for Europe's space research and technology was a white building in a Parisian side street near the École Militaire. Ancient enemies met in peace at the headquarters of the European Space Agency, where national delegates conjured from official paperwork their shared adventures in the border-free vacuum above the clouds. Scientists and engineers trooped to the Rue Mario-Nikis desperately hoping for approval of their missions.

The science programme committee met there on 8 July 1980 to judge whether Giotto could, after all, squeeze in ahead of the cherished Hipparcos mission. The director general said that no firm offer of a launcher had come from NASA, so any commitment to Giotto had to be based on the full cost of Europe going alone, using Ariane. Dale reported 'no insuperable problems' in the design or operation of the spacecraft. He could deliver the mission for 83 million accounting units, scarcely above the ceiling specified in March.

The national delegates were almost unanimous in making Giotto the next scientific mission of the agency. Only France voted against it, because of the delay to Hipparcos. The door was not to be slammed on possible NASA participation, but any switch to an American launcher would require a fresh decision by the science programme committee.

Europe thus made up its mind to go to Halley's Comet unaided. Dale achieved his wish to become Giotto's project manager, and Reinhard was the natural choice for the project scientist. Five frantic months of selling the idea were over. The certainty that Giotto would depart for Halley in July 1985, or not at all, promised five frantic years ahead.

That a comet had a solid nucleus was only a hypothesis in 1980. A few experts still clung to the previous notion of a comet as a swarm of grains on similar orbits. When the Giotto mission was announced, Raymond Lyttleton of Cambridge scoffed at its task of finding the nucleus of Halley's Comet.

'The idea has been able to survive for several years,' Lyttleton wrote in a London newspaper, 'precisely because such a postulated nucleus would be too small to be detectable from the Earth.'

But not too small to be seen by a visiting spacecraft. Hypotheses in science were confirmed or disproved, not by the numbers or the authority of their adherents, nor by the elegance of the arguments, but by evidence. The time was ripe for comet science to come of age. Even for those convinced of the reality of the nucleus, its character was poorly defined.

Uwe Keller was already assembling a team to create the camera with which Giotto would look for the Halley nucleus. He had been unenthusiastically associated with the camera for the defunct Halley probe, but as an instigator of Giotto he had to make sure that the concept was carried over into the new spacecraft.

'It is like riding on a carousel and trying to look at one particular star through a telescope, once in every revolution.' That was how Keller described the task of observing the nucleus of the comet from a spinning spacecraft.

The solution that Alan Delamere of Ball Aerospace in Colorado had offered for the Halley probe depended on solid-state light detectors called charge-coupled devices, or CCDs. They were novelties, at least for civilians, and combined a good response to light with a capacity to store the visual information. Delamere's idea was to expose the whole image quickly while the comet was in the camera's field of view, and then to spend several seconds reading the information from the array of CCDs, while the spacecraft's rotation was sweeping the camera across the wrong parts of the sky. Although simple in principle, the technique depended on control systems of exquisite complexity. No one had ever built such a camera.

In the spring of 1980, Keller considered the simplest option, of paying the Ball company to develop the instrument. But with whose money? Although the European Space Agency provided a spacecraft, it relied on the national organizations that funded Europe's scientists to supply the instruments. Why should the German government, for example, pay for a US-made camera? Instead, Keller sounded out Delamere and Ball Aerospace about whether they would help a European team build the camera. They agreed to do so.

Playing the impresario, Keller canvassed colleagues around Europe looking for someone who would take charge of the camera as the 'principal investigator'. That implied heavy financial obligations, and he found no takers. The awful truth dawned on Keller that he would have to do the job himself. He was an astrophysicist, not a telescope-maker or electronic engineer, and despite some experience with space instruments he had never handled anything so elaborate. But the Max-Planck-Gesellschaft and the Bonn government were willing to back him with funds, as the German tradition in comet science paid a dividend.

Keller combed Europe again for 'co-investigators'. So far from making the camera team a Germans-only club, he wanted all the scientific, technical and financial backing he could find in other countries. From France, Jean-Loup Bertaux of Verrières-le-Buisson was an obvious recruit. Like Keller, he had been a member of the original camera team for the International Comet Mission, and he gladly joined the new effort. So did a group at Marseille, Italian and Belgian physicists at Padua and Liège, and some individual British scientists. By August 1980, Keller was rushing to meet the deadline for submitting his camera proposal to the European Space Agency. Delamere of Ball Aerospace and Rüdeger Reinhard from Noordwijk helped with the technical description.

As he had tried to involve everyone he thought was interested, Keller was startled to hear of a competing bid for the Giotto camera. It came from the French scientist who first introduced Keller to space research ten years earlier, Jacques Blamont. He knew about Keller's camera, because Bertaux was a researcher in his laboratory at Verrières-le-Buisson, and he doubted Europe's ability to master the necessary technologies in time for the encounter with Halley.

During that summer of 1980, Blamont was a 'distinguished visiting scientist' at the Jet Propulsion Laboratory in Pasadena. He offered JPL a deal. If the American lab would devise a camera for Giotto, he would sponsor its adoption. Although the camera would be built in the US, the rules required that a European scientist be nominated as principal investigator. Blamont offered himself in that role, with Bertaux as his deputy.

JPL's unrivalled skill in space imagery culminated with the Voyagers launched on their grand tour of the outer Solar System. Scientists and the public alike had gasped at the images of Jupiter and its moons that came back from the Voyagers in 1979. Anyone betting on the prowess of Pasadena versus Katlenburg-Lindau would not have to think long.

Blamont himself was one of Europe's top space scientists. After France's Centre National d'Études Spatiales was created in 1962, Blamont served for ten years as its first technical and scientific director. He was a member of the exclusive Académie des Sciences in Paris and a foreign associate of the US National Academy of Sciences. Most impressively in the context of Giotto, Blamont was co-chairman of the scientific committee planning a Soviet-led Vega mission to Venus and Halley.

Keller had a fight on his hands and his team was falling apart. Bertaux had to pull out, as his name was on the rival bid, but the withdrawal of the Marseille group was painful. In theory, instruments for the European Space Agency's missions were chosen purely for their scientific and technical merit, but in the real world national sensibilities could not be ignored entirely. French delegates, in the Paris offices of the agency, would be keen for Blamont to win the camera slot. They could argue that Giotto's instrument deck was likely to be loaded with much German equipment, notably the chemical sensors.

A French ally therefore seemed essential to Keller, and he thought of Roger Bonnet. Although a former student of Blamont, he broke away during the students' revolt of 1968, when discontent swept through the universities and institutes of France. Bonnet set up a rival laboratory of stellar and planetary physics at Verrières-le-Buisson. Keller had worked with him in the US on a NASA project.

'Listen, I'm going to talk to you very openly,' Keller said to Bonnet when he reached him by phone and described his predicament. 'I think we need French support in this team and I'm asking you, don't you want to come in? You have big laboratory. Can't you make a contribution? What about the optics?'

'Let me think about it,' Bonnet said. 'I'll call you back in two or three days.'

Bonnet was angry about Blamont's proposal. He saw as clearly as Keller how it could be presented as a French versus German issue, yet in Bonnet's opinion it was a case of the US versus Europe, and Blamont's camera was a Trojan Horse. If JPL generated the images of Halley's nucleus, it would reap the chief glory from the Giotto mission. Bonnet assessed his lab's resources. It was already committed to making infra-red instruments for the Soviet Vega spacecraft, but he decided he could fit in a responsibility for the optical system of Giotto's camera. He assured Keller that he would not only join the camera team but take a practical part.

Keller bolstered the prestige of his team by recruiting two eminent comet scientists to his list of co-investigators. Ludwig Biermann was the easier catch, as Keller's former teacher, but Fred Whipple from the US also joined in. For a big name in European space science, to balance Blamont himself, Keller persuaded Giuseppe Colombo from Padua to join the team.

Blamont's instrument had a neater name: CHIC (Comet Halley Imaging Camera) as opposed to Keller's HMC (Halley Multicolour Camera). They solved the problem of the spinning spacecraft in broadly similar ways, but a scanning mirror would direct the light into Blamont's camera, while the whole camera would revolve in Keller's scheme. Blamont's could not look backwards at the nucleus after the encounter, as Keller's could, but Blamont reasoned that Giotto would probably be damaged by dust before it passed the nucleus.

The rivals had to go through selection procedures, as Giotto itself had done earlier in the year. During November 1980 independent referees commented on the scientific merits of the proposals, while at Noordwijk an ESTEC team checked their compatibility with the spacecraft. That month, the JPL camera in Voyager-1 was sending pictures from Saturn of astonishing beauty and detail.

The Solar System working group met at the agency's headquarters on 15 and 16 December, to review proposals for Giotto experiments received from laboratories around Europe. Fourteen scientists from seven countries set out to pick the instruments that would serve the mission best. Although the camera was only one item among a dozen, the group knew that it was politically the most sensitive.

Its members were open to persuasion by technical arguments, and understood the arithmetic of costs and weights. As intelligent men aware of the delicate multinational nature of the mission they could be cajoled by hints about 'fair play', but they could not be browbeaten. Blamont's manner won him no friends. He plainly disliked having to explain himself and his splendid instrument to scientists ten or twenty years younger than himself, and he attacked Keller's scheme.

'We prefer to ensure the maximum scientific return and overall success of the Giotto mission,' Blamont declared, 'not to take advantage of this project to painfully acquire, in Europe, competence in the field of CCD imaging – at the risk of meeting desperate difficulties with the development schedule.'

The group considered and rejected a shotgun marriage between Blamont and Keller, before deciding by a narrow vote in favour of Keller. The science advisory committee, chaired by Klaus Pinkau, endorsed the choice a few days later.

Blamont complained of bias. In a letter to the director general of the agency, he cited the participation of Rüdeger Reinhard, a co-investigator for the Keller camera, in the ESTEC assessment team. And although scientists whose proposals were under discussion were supposed to leave the room, those present during the decisive hours in the Solar System working group included a British member of the Keller team, and a scientist from Keller's institute at Katlenburg-Lindau. Ernst Trendelenburg from the agency's headquarters had intervened with a telex from NASA indicating that American-made CCDs would be available to any European group, a move which in Blamont's opinion obscured the issue of expertise. Pinkau's advisory committee could, he said, only rubber-stamp the working group's verdict.

The letter was written in English but Blamont referred to Herr Keller, Herr Trendelenburg, Herr Reinhard and Herr Pinkau, as if to emphasise their nationality. He also warned that a satisfactory camera could be accomplished only by experienced people, not by 'amateurs'. Blamont predicted that Keller's team would soon have to adopt the essential ingredients of his own proposal, in which case he wanted to be put in charge of the experiment.

Blamont always had a strong case, both technically and politically, and his misgivings about the prospects for the Keller camera were not ill-founded. Yet in arriving at their decision most of those involved shared Bonnet's instinct. Why go all the way to Halley's Comet so that the Americans could take the pictures?

That the Americans were in a tizzy, after the collapse of their grand mission to Halley and Comet Tempel 2, was the most charitable explanation of their odd behaviour throughout 1980. Contradictory signals of friendship, arrogance and disparagement perplexed officials of the European Space Agency who were under orders to try to make a deal with NASA over Giotto. Eventually the bigheartedness of the Americans would shine through, but not soon enough to give them any real say in the mission.

A private US company, RCA, expressed its horror at the loss of national prestige by offering to build a Halley spacecraft itself, if NASA would launch it. In another venture, an organization called the Halley Fund invited donations from the public to pay for a mission. These proposals came to nothing but a veritable fleet of other people's spacecraft, lining up to fly by everyone's favourite comet, plainly bruised American egos.

The Soviets had the most elaborate plan. They intended to launch two spacecraft towards Venus, in 1984, and would send one or both of them onwards from Venus to an encounter with Halley. The Japanese (raw beginners!) were completing a new launcher for their mission, although they gave up an interesting idea of catching Halley on the far side of the Sun, when it was out of sight from the Earth. And the Europeans were preparing to send their spacecraft much deeper into the head of the comet than anyone else's.

But Giotto was no good, according to Joseph Veverka of Cornell. He was the chairman of NASA's comet science working group who had so quickly dismissed the idea of an American flyby of Halley when the International Comet Mission collapsed. He opposed any direct US involvement in Giotto, saying it could not deliver the required results.

'You will go zooming off into nowhere,' Veverka affirmed, in commenting on a flyby of the Giotto type. 'The situation will be a disaster.'

When a flyby was plainly the only option left to the US, Veverka changed his tune, to say that only NASA could execute it properly. Paul Weissman of the Jet Propulsion Laboratory declared that a US flyby mission 'would do the hard science that the European, Japanese and Russian missions cannot do'. American arrogance was becoming insufferable.

Yet when Rüdeger Reinhard described the Giotto mission in all its detail to open-minded American colleagues, he convinced them that they could do little in an ordinary flyby mission that would improve significantly on Giotto. They came up with a complementary proposal to scoop up samples of dust and bring them to the Earth for thorough analysis. The spacecraft would have wing-like panels that would open in Halley's atmosphere and then fold again, to protect the collected material for the return journey.

But JPL considered that the US taxpayers wanted instant images, not long-term chemistry. Apart from its bid through Jacques Blamont to supply the camera for Giotto, JPL sought funding for Voyager-like spacecraft to go to Halley and, as it claimed, take much better pictures than Giotto could ever obtain. As the price-tag was about twice Giotto's, NASA headquarters thought it poor value for money. JPL held out, hoping for better treatment if Ronald Reagan won the presidential election late in 1980.

While internal wrangling went on, NASA lost its chance to be a partner in Giotto. Relations between the two agencies were strained by an American cutback in an existing joint project, for which Europe was already building the Ulysses spacecraft. While officials made many transatlantic flights by Concorde to keep that mission alive, the NASA administrator let the Giotto decision day go by, in July 1980, without making any proposal to Paris about an American launcher.

Three months after the Europeans had formally resolved to use their own Ariane, NASA said it might be able to supply a Delta launcher for Giotto, as well as helping with communications and the camera. In exchange, US scientists would take charge of two experiments in Giotto. Europe's scientists and officials duly considered the proposal, and in diplomatic language told NASA to get lost.

President Reagan was no more inclined than President Carter to fund an expensive mission to Halley. The Americans would not even have a flyby. The scientific case for a 'mere' flyby had not been developed in the US, as it was by Johannes Geiss and Hugo Fechtig in Europe and their counterparts in Japan and the Soviet Union. The real loser was the human species in its efforts to understand its place in the Universe. A different kind of mission, like the dust-sample return, could have been a fine addition to the international fleet going to Halley.

A few experiments were put together in the US to observe Halley from Earth orbit, but the teams at JPL and elsewhere hoping for a US comet mission found themselves mainly relegated to the role of passive gurus. JPL became the headquarters for the International Halley Watch, which performed the vital though unspectacular task of coordinating observations of the comet from the ground and from space. And JPL's scientific rivals within NASA, at the Goddard Space Flight Center in Maryland, cunningly directed an existing spacecraft to reach another comet months before the international fleet arrived at Halley. That applied a plaster of sorts to injured American pride.

At the laboratory level, where the real science was done, everything looked far healthier. Quickfooted Americans became co-investigators in experiments in the European and Soviet missions. More than forty of the 160 or so scientists working in the Giotto experimental teams were from the US – a greater number than from any one European country. They often had valuable know-how.

When tempers cooled, and NASA realized that Giotto was really going to happen, David Dale gained excellent help from JPL's Deep Space Network of tracking stations at a sensible price. He confirmed that the Australian telescope at Parkes was still the primary station for the Halley encounter but American support was reassuring. Later, a special operation evolved in which the American ground stations helped to guide Giotto with surgical precision into the heart of the comet.

But for the lure of Halley's Comet the chances of any comet mission flying in the 1980s would have been slim. Had the Americans set up a comet flyby of their own, immediately after the demise of the International Comet Mission, there would have been scant enthusiasm for Giotto. And in years to come, over a beer, Europe's space scientists would sometimes wonder whether their mission to Halley could ever have happened without the timely intervention of Ernst Trendelenburg's mother.

CHAPTER THREE

THE RIGHT SHAPE FOR GIOTTO

THE FAIRY-TALE castle of Smolenicie in Czechoslovakia was the venue for an international conference on solar-terrestrial physics in the summer of 1980. Susan McKenna-Lawlor was there, a petite and bonny Irish professor who dressed in primary colours and saw the world in the same intense way. She sighed with relief as she sat down to dinner with her fellow-scientists in the banqueting hall. After unexpected difficulties in crossing the Iron Curtain to attend the meeting, she had arrived in time to present her paper on solar flares. Now she could relax.

Sitting next to her was a tall young man with a boyish smile. 'I'm Rüdeger Reinhard,' he said. 'I'm the project scientist for the Giotto mission.' Their conversation wandered amiably and the meal was almost over before Reinhard came out with a surprising question.

'Ireland is now a full member of the European Space Agency.' he said. 'So why is there no Irish experiment proposed for the comet mission?'

McKenna-Lawlor felt the shock of a moment of destiny. Once before in her scholarly life she had had a similar experience, as a university student. At her school for 'young ladies' she had learnt virtually no science and seemed set on a musical career, but arriving at University College Dublin she decided to sample the science. The quantum theory of the hydrogen atom moved her deeply. It was like poetry, and Susan McKenna decided there and then to be a physicist.

Despite a late start, she did well and proceeded to postgraduate research at the Dublin Institute for Advanced Studies. She chose solar flares, explosions on the Sun, for her doctorate at the University of Michigan, and while in the US she lectured NASA's astronauts on the dangers to space travellers of energetic particles from solar flares. Her association with NASA continued when she returned to Ireland, through various space missions for monitoring activity on the Sun.

McKenna-Lawlor therefore knew enough about the technicalities of spaceflight to think that an Irish-led experiment for the Giotto mission to Halley's Comet would be very difficult to mount. Her country did not possess the clean rooms, environmental testing equipment, particle accelerators for calibration, or the special hardware for preparing a flight-qualified instrument.

Despite that, she felt inspired to look Reinhard in the eye and say, 'Why not?'

'I advise you to visit the Max-Planck-Institut für Aeronomie,' he said. 'They are planning several experiments for Giotto there and would probably help with technical facilities. Talk to the director, Ian Axford. But you must be quick – your detailed proposals have to be in by the middle of October.'

At Katlenburg-Lindau, Axford and McKenna-Lawlor discussed her concept for a particle telescope with semiconducting detectors. It would register energetic particles from the Sun during Giotto's cruise to Halley. At the comet itself it would detect other energetic particles accelerated locally. Axford offered the use of a clean room and testing equipment until she could install facilities of her own in Ireland. Erhard Kirsch would join the team from the German institute, as co-investigator, and provide engineering support.

'You need a name for it,' Axford said, 'representative of Ireland.'

McKenna-Lawlor ruminated. 'EPONA, that's it. Energetic particles onset . . . what word? . . . admonitor.' She explained that Epona was also the beautiful and mysterious Celtic goddess associated with the Solar Year.

She hurried back to Ireland to recruit participants from the Dublin Institute for Advanced Studies and to tackle a government agency for funds. Her own base was at St Patrick's College, a famous seminary standing beside the ruins of a 12th-Century castle in the small town of Maynooth, out on the Galway road. There McKenna-Lawlor startled her clerical and literary colleagues with the news that they were going into space technology.

Her particles fell outside the range of energies specified for the 'model payload', which Giotto inherited from the International Comet Mission. This established a clear order of priority, and put McKenna-Lawlor's proposal in Category 3 – experiments of an unsolicited kind with little hope of incorporation in Giotto. She was in for a roller-coaster ride of hope and disappointment.

Indeed, McKenna-Lawlor had little idea at first of the wheeling, dealing and rivalry between the larger academic space laboratories in Europe that were vying for a place in Giotto. Apart from the disputed camera, the top requirement was for mass spectrometers to analyse the comet's chemical composition. The Germans would be fighting among themselves over these Category 1 instruments, although the Swiss seemed to have a clear run with the ion mass spectrometer for identifying charged atoms and molecules.

Shrewd experimenters in France, Italy and the United Kingdom saw their best chances in a second category of instruments which included dust-impact recorders and ultraviolet analysers for the comet's gas. Sensors for magnetic fields and subatomic particles, which would examine the collision between the comet and the solar wind, were also in this category. Successful play in the acceptance game meant forming international alliances. These reduced the number of competing proposals and strengthened those that went forward.

Alan Johnstone had been thinking for several years about Halley. He was a black-bearded physicist from the Mullard Space Science Laboratory, an outpost of London University at Holmbury St Mary in the wooded hills of Surrey. Johnstone had agreed in 1978 to work with colleagues in Texas on a joint proposal for any mission, US or European, that might be going to the comet. It would be an instrument, or set of instruments, to detect the subatomic particles of the solar wind and see how they changed in strength and direction as they were joined by particles originating from the comet itself.

Assuming that German groups in Munich and Katlenburg-Lindau would be planning a similar proposal, Johnstone looked for a wider European partnership. He threw in his lot with a strong Italian group from an interplanetary physics laboratory at Frascati. He also contacted Swedish friends at the Kiruna Geophysical Institute, and strengthened his British base with co-investigators from Chilton and Cardiff. Then, to Johnstone's surprise, the German contenders wanted to join in the same proposal.

In the early stages, a question was left open. Who would be the principal investigator, responsible for submitting the proposal and supervising the execution of the experiment if it should be accepted? Eventually the Italians asked Johnstone to do it, and he found himself leading a five-nation alliance, with co-investigators from Germany, Italy, Sweden, the United Kingdom and the United States.

Everything to do with the mission was a matter of teamwork. If the author named all of the hundreds of individuals and dozens of organizations involved in the science, engineering and operations of Giotto who deserve a mention, this book would read like a telephone directory. The reader is asked to take it for granted that everyone named was part of a larger team.

The principal investigators of the various experiments had to exercise leadership before, during and after the mission, and to represent their teams in all dealings with the project management and mission control. It was not unfair that this multinational project to measure subatomic particles came to be called the Johnstone plasma analyser.

'I take the blame and the credit,' Johnstone said.

While Uwe Keller and Jacques Blamont, contenders for the camera, were the angry gladiators when the European Space Agency's Solar System working group selected the instruments for Giotto, other proposers sat or paced about like rival job applicants, glum or cheery. As the process dragged on for two days, at the white building in Rue Mario-Nikis, they could size up their possible companions in the mission to Halley's Comet.

Alan Johnstone was striking a bargain with Henri Rème, a tall Frenchman from Toulouse, who was proposing another detector for subatomic particles, with impressive support from the University of California. Among other would-be principal investigators in Blamont's entourage, Susan McKenna-Lawlor met the jaunty Jean-Loup Bertaux and a bright Parisienne, Anny-Chantal Levasseur- Regourd. 'Call me Chantal,' she said.

Like McKenna-Lawlor, Levasseur-Regourd wanted to slip a small, lightweight instrument into Giotto of a kind not visualized in the model payload. She called it HOPE, for Halley Optical Probe Experiment. It would be a telescope looking backwards along the spacecraft's flight path through the comet. Filters would pick out light scattered by dust in three bands of wavelengths, and characteristic light emitted by four kinds of molecular fragments. The increase in intensity, as Giotto penetrated deeper into the comet, would measure the abundance of material in the spacecraft's wake.

Hans Balsiger from Bern, as neat and bespectacled as a Swiss bank manager, was friendly with everyone. He came from the laboratory of Johannes Geiss, who had long battled for the comet mission, and Balsiger knew that his ion mass spectrometer was a certainty. Equally confident, but more aloof with it, was the crew-cut Jochen Kissel from Heidelberg, built like an American footballer. His dust mass spectrometer for analysing vaporized dust grains in the comet was earmarked not only for Giotto but the Soviet Halley mission too.

So much time and emotion went into the arguments about the camera, that the Solar System working group was unable to finish its selection of the other experiments. As ESTEC and independent referees had vetted them all, approving experiments in the model payload which were met by single proposals was almost a formality. Besides Balsiger's and Kissel's spectrometers, the jury quickly adopted a system of dust-impact counters from Tony McDonnell of Kent University at Canterbury in England, and magnetic instruments from Fritz Neubauer of Cologne in Germany. Both detectors of subatomic particles in the interplanetary plasma, Johnstone's and Rème's, were selected because they had arranged to share the allotted mass between them. A small chemical instrument from Axel Korth of Katlenburg-Lindau joined Giotto as a component of Rème's proposal.

Hard choices began with teams led by Dieter Krankowsky of Heidelberg and Erhardt Keppler of Katlenburg-Lindau, offering a neutral mass spectrometer to measure the masses of uncharged atoms and molecules in Halley's atmosphere. Krankowsky's instrument was much the heavier of the two and magnetic enough to interfere with measurements of magnetic fields in space. The Solar System working group nevertheless favoured it on the grounds of likely performance.

Neither of two proposals for an ultraviolet sensor quite fitted the specification in the model payload. One came from the Rutherford-Appleton Laboratory in England, and the other from Bertaux of Verrières-le-Buisson. The group considered Levasseur-Regourd's optical experiment jointly with these, and put Bertaux's experiment at the bottom of the list. This doomed to exclusion a scientist who had fought for the comet mission back in 1978, worked on the old Halley probe, and switched from Keller's team to Blamont's for the camera proposal. Bertaux's hopes, like Blamont's own, now lay in Moscow with the Soviet mission.

With time running out, the jury could thankfully leave aside radio experiments that required no instruments aboard the spacecraft. A plasma-wave experiment, to detect certain interactions between the solar wind and the comet better than the Johnstone-Rème experiments, was proposed by a group of European Space Agency scientists at Noordwijk. It failed to win a place because it might need booms protruding from the spacecraft. That left Susan McKenna-Lawlor's energetic-particles experiment. The Solar System working group declared that it should be considered for inclusion, but gave it a low priority.

The meeting broke up with tempers frayed and the total weight of approved instruments still exceeding by twelve kilograms the fifty-three kilograms allocated to them. The science advisory committee, to which the Solar System working group reported, was therefore left to whittle the payload further. To save weight the committee eliminated not just Bertaux's ultraviolet instrument but also the British offering, partly on the grounds that a Japanese spacecraft would observe Halley by ultraviolet light. It kept Levasseur-Regourd's instrument aboard but it struck Neubauer's magnetometer off the list. McKenna-Lawlor's energetic-particles experiment was discarded too.

'Don't give up hope, Susan,' Reinhard advised her when the bad news came through. He explained that changes to the spacecraft, already under discussion, might still make room for Neubauer and herself. And in the spring of 1981, the magnetometer and the energetic-particles instrument were restored to the Giotto payload. There was widespread pleasure in the European Space Agency that, alongside four experiments from Germany, two from France, two from the United Kingdom and one from Switzerland, an Irish experiment would go to Halley's Comet.

The instruments were an odd collection of boxes and tubes. McKenna-Lawlor's EPONA was the smallest and Keller's HMC the largest. When the spacecraft engineers designed them into Giotto, most went on the instrument deck behind the bumper shield. They would peep cautiously out at the comet, around the edge of the shield, presenting as small a target as possible to the comet's dust. What power and control systems they would use, and how their data would emerge for transmission to the Earth, were other matters for attention. Once they were settled, a scientist who sought to change anything was asking for an argument.

And experimenters who might be gratified at first by the interest shown in their work, not only by Reinhard but by the project manager David Dale and his engineers, gradually came to realize that the big brothers were watching them. Indeed, Dale believed in maintaining creative tension between his team and the scientists. He needed their instruments made, tested and delivered on time and to specification, yet he had no direct authority over them, other than the ultimate sanction of leaving an experiment behind when Giotto went to Halley's Comet.

The split method of funding the mission made difficulties for Dale. Unlike NASA, which paid for a spacecraft's instruments, the European Space Agency provided the spacecraft but left the scientists and their national funding agencies to pick up the bills for the experiments. When it came to managing the scientists, that weakened Dale's position without sparing him all financial risk. The agency delivered the instruments to the industrial contractor building the spacecraft. If they did not fit, or were late, or failed a test, the contractor billed the agency for the extra costs arising. Dale never forgot that Giotto was advertised as a low-cost project, and every penny counted.

So did every gram. A prime duty of any experimenter, in Dale's opinion, was to keep within the authorized mass. The stinginess of the project manager in this regard became legendary at the outset, when an error in a committee document assigned less mass to one experiment than it should have done. The scientists concerned said that they could not possibly make an instrument so light. Dale insisted that the official figure had to stand. The discrepancy was 250 grams, just one 4000th part of the total mass of the spacecraft, yet it became the subject of a big dispute. Dale lost in the end as he knew he would, when the scientists appealed to Paris, but he had made clear what a resolute weightwatcher he would be.

The city of Bristol prospered by innovation. From this western port, five years after Columbus, England's first official expedition left for the New World, and the townsfolk later grew rich on sugar and the slave trade. In the 20th Century it pinned its hopes on aviation. From the runway at Filton in the northern suburbs, machines of every generation climbed over the estuary of the Severn, from Bristol fighters of the First World War to the British Aircraft Corporation's supersonic Concorde.

By 1980 the company was called British Aerospace. In an enlargement of the collaboration with the French that created Concorde, Europe's Airbus was setting out to challenge Boeing in the civil aviation market, using wings made in Bristol. And British Aerospace was earning a name for itself in the satellite-building business, in spite of London's notorious indifference to spaceflight.

The latest task was to adapt one British Aerospace spacecraft, Geos, into another, Giotto. David Link was the lean-faced, soft-spoken project manager at Bristol, an electronics engineer by origin. Others in his team had trained as physicists. In the early decades of the Space Age a formal education in spacecraft engineering was hard to find. You learned it by doing it, as Link had done.

He shared David Dale's belief that the interception of Halley's Comet was the chance of a lifetime. It could never be a money-spinner for the company, but a leading part in Europe's first venture into deep space could win it some headlines. And as an engineering task it was fascinating.

During the study phase, before the final decision in July 1980 to go ahead with the mission, Link and his engineers had sketched a design for the spacecraft based on Geos. The original Geos was a drum-shaped satellite built around an onboard rocket motor. Solar cells decorated the curved face of the drum to convert sunlight into electricity. The satellite and its motor spun around their common axis to make the whole spacecraft into a gyroscope, thus stabilizing it. Protruding booms carried some of the scientific instruments. How did Geos have to change, to become Giotto?

The design of any spacecraft depended on the special requirements of the mission, and in the case of Giotto these began and finished with the encounter with Halley. Computable facts of astronomy governed the scenario during those crucial few hours. The comet's orbit slanted in relation to the Earth's, and an interception when the comet crossed the Earth's orbit, in March 1986, required the least rocket power. Six years ahead of the event, the European Space Agency's flight dynamicists calculated Giotto's course and speed in its approach to the comet, and the bearings of the Sun and the Earth at the time of the encounter.

The engineers' first concern was with Halley's dust. At 68.4 kilometres per second, Giotto's speed relative to the comet would be even faster than that visualized for the defunct Halley probe, and high-speed dust impacts would destroy an unprotected spacecraft. An immediate change from Geos to Giotto was to strip off the booms, because impacts on them could too easily destabilize the spacecraft.

Giotto would travel into the comet with the nozzle of its spent motor facing the onslaught of high-speed dust. The nozzle had to be sealed with a lid to prevent dust piercing the spacecraft by that route. Another shield, making a ring around the motor, would protect the scientific instruments and the rest of the spacecraft. As the shield was expected to be wider than the drum, a fairing around the instrument deck flared like a skirt towards the shield.

The dual bumper shield, invented by Fred Whipple and adapted by Rüdeger Reinhard for the old Halley probe, was carried over into Giotto, with variations. The gap between the thin front sheet and the thick rear sheet was slightly narrower. And the material for the rear sheet, designed to absorb the scattered fragments of large dust grains penetrating the front sheet, would be Kevlar, a synthetic textile of para-aramid fibres much favoured by soldiers and policemen for body-armour. Used as reinforcement for an epoxy plastic, it made a tough composite.

The suggestion came from Robert Lainé, the French engineer in David Dale's team at ESTEC who played a prominent part in Giotto's evolution. In his opinion the added protection of Kevlar outweighed Reinhard's worries about introducing carbon compounds that would pollute the spacecraft's environment during the encounter. And so Giotto acquired its bullet-proof vest.

On the other end of the drum, least exposed to the comet, went a radio dish, Giotto's main link with the Earth. The only time when the spacecraft could not be slewed to suit communications needs was during the encounter with Halley. Giotto was therefore born with the permanent squint appropriate to a dedicated comet raider. Its radio dish would always be tilted at just the angle required for its beam to find the Earth in its computed direction at the time of the encounter, while Giotto hurtled, bumper shield first, along its prescribed track into the comet. Measured from the spin axis, the beam had to go off at 44.2 degrees.

The spin of the spacecraft would swing the beam out of alignment with the Earth, unless a despin motor in the mounting of the dish compensated exactly for the rotation. A mast in the middle of the drum's 'safe' end carried the main transmitters, a pair of microwave amplifiers called travelling wave tubes, which fired the signal pulses into the radio dish, like the bulb shining into the reflector of a car headlamp. The mast also had a secondary, look-anywhere antenna for use in circumstances, hopefully deliberate, when the dish was not trained on the Earth.

Diagrams of this conceptual adaptation of Geos appeared in the documents on which the science programme committee in Paris based their authorization of the Giotto mission. To Link and his team it was the natural basis for the detailed design that British Aerospace would propose to the European Space Agency. Indeed they saw strong reasons for sticking to this approach.

The idea of adapting Geos to a low-cost comet mission was the very genesis of Giotto. British Aerospace, Giuseppe Colombo and Ernst Trendelenburg vied for the credit for it. As the agency's director of science, who claimed the idea came to him in his bath, Trendelenburg kept saying, 'There is no need to reinvent the wheel.' The selection of British Aerospace as prime contractor without the usual open competition had been entirely due to its affiliation to the previous spacecraft, and eyebrows might rise if Giotto did not resemble Geos. Finally, questions of schedule and budget encouraged the maximum use of a proven design.

When the team of engineers at Bristol set about refining the Geos adaptation, they saw difficulties with the power supplies. The size of the Geos drum defined the area of the solar cells. At Halley these would generate 141 watts of electricity, compared with 189 watts needed when the experiments and the main transmitter were running. A battery was the answer, sufficient to supply ample power for a couple of hours even if the comet dust annihilated the solar cells. Temperature control within the spacecraft was another problem. The sunlight would vary in intensity by a factor of two during the cruise to Halley, and would fall unevenly on different parts of the spacecraft, depending on its attitude. The British Aerospace team visualized the use of phase-change materials that would absorb or release heat as the temperature rose or fell.

Working a hundred-hour week, the engineers attended to a hundred matters ranging from the elements of electronic control to the mechanical feasibility of the despin motor for the radio dish. They explored the allocation of work to subcontractors across Europe, and obtained bids. Keeping costs within the set budget, while restricting the mass of the spacecraft to 840 kilograms, gave the engineers an acute sense of squeezing a quart into a pint pot. In March 1981, after eight months' work, British Aerospace formally tendered its design for Giotto to the European Space Agency.

Dale and Lainé in the project office, other engineers from the agency and a string of outside experts examined every aspect of the technical design and the management proposals. They awarded marks under various headings. Their evaluation went to the agency's industrial policy committee which consisted, like the science programme committee, of delegates from the member states. Totted up, the total of marks was too low. The committee rejected the tender from Bristol.

British Aerospace demanded an explanation, and David Link and his team attended a grim-faced meeting at Noordwijk. Independent experts reported that the power supplies and the arrangements for controlling temperatures in the spacecraft were inadequate for the mission. Special problems with power and heat would arise when parts of the spacecraft cast a shadow on the solar cells. Complaints about too much reliance on the Geos spacecraft, made the Bristol team think that the client had moved the goalposts. Link suspected that Dale had encouraged this new and surprising aversion to Geos.

After the meeting Link went up to him and asked, 'What do we do now?'

'Well why don't you come round to my house and we'll have a beer?'

Giotto was in crisis. British Aerospace was in danger of losing the contract. With much time apparently wasted, the European Space Agency's mission to Halley's Comet could now fail to reach the launch pad by the due date. The project managers whose careers were in jeopardy sat in Dale's conservatory at Voorschoten, in the suburbs of The Hague. Till one o'clock in the morning, the two Daves swapped jokes about the vagaries and mishaps of space projects. Dale went to bed convinced that Link was a manager he could trust, and whose commitment to Giotto was total. Although they were officially at loggerheads, the personal bond between the men might survive the brutal weeks ahead.

While David Link returned to Bristol to explain the grievous state of affairs to his own management, David Dale and his project team at ESTEC took stock. As Link correctly guessed, they had fallen out of love with Geos several months earlier. Robert Lainé had worked on the Giotto design independently of British Aerospace and conceived a markedly different spacecraft. He too had done the sums about power and temperature control and thought that Giotto's drum should be altogether bigger than Geos's. Lainé persuaded Dale that to stretch everything on Geos was taking too many risks.

Even financially the bonus from Geos was an illusion, Dale concluded. Companies were telling him, 'We don't build that technology any more. If you want us to go back to Geos electronics we'll have to retrain people. We can build the old Geos structure, certainly, but a new structure would be cheaper.' The state of the art in subsystems was represented better in other spacecraft under development for the European Space Agency. Ulysses, for one, was intended for a deep space mission to the Sun and had similar communications needs. The same microwave transmitters and receivers would do for Giotto.

'Rather than insisting that Giotto will be a Geos derivative with all our new bits and pieces hanging on it,' Dale said to his team, 'why don't we just say we're going to use the maximum amount of existing technology?'

Yet the Geos connection had seemed a bright idea just a year before. It enabled the British Aerospace engineers to begin work far sooner than if the study and design contracts had gone out to competitive tender. And had the project started with a blank sheet of paper, the result might have been an exotic design too novel to fulfil within the available time, never mind within the budget. The existing design was judged imperfect, but at least it was buildable. It addressed the right questions, and the only sensible way forward, Dale decided, was to set about improving the answers.

Who should do it? Dale had to make official suggestions in the aftermath of the failed bid. One was that ESTEC itself could design the spacecraft and find the subcontractors. Dale also went through the motions of recruiting an Italian company as industrial architect. But he was not sorry when his political masters threw out these possibilities. Persuaded by the British delegation, and having an eye to the time factor, the industrial policy committee redirected Dale to Bristol. He was to set the project back on track with British Aerospace remaining the prime contractor.

The project engineers at ESTEC and British Aerospace then worked in tandem for nearly six months. Leadership gradually reverted from the Europeans to the British. Dale and Link agreed on stern rules. If any pair of engineers from the two sides failed to harmonize their ideas both would be removed from the project. So would anyone who harped on the differences of opinion revealed in the review of the original tender.

The design for Giotto which emerged was very different from the Geos adaptation. The central mast carrying the radio transmitters gave way to a tripod with crooked legs. More significantly, the drum could be bigger and heavier, thanks to a change in the launch scenario, and a more powerful onboard motor available from France. With the drum almost as broad as the bumper shield, the flared skirt enclosing the scientific experiments became almost straight. To a casual glance, the new Giotto looked like a true cylinder. The total height of the spacecraft was 2.85 metres, and its diameter at the bumper shield was 1.87 metres.

The solar cells surrounding the bigger drum would give 190 watts or more at Halley's Comet rather than 141. Four batteries using fifty-six silver-cadmium cells would boost the power available during the encounter. The engineers tackled afresh the vexed question of temperature control. Thermal blankets would protect Giotto's electronic systems and scientific instruments, while radiating and reflecting surfaces with adjustable shutters would provide cooling. Key regions could be heated electrically.

Everyone's thoughts focused on the spacecraft's performance during the critical hours inside the head of Halley's Comet. As there were doubts about Giotto's survival, weight and power could be saved if the spacecraft had no general recording facility. Data from the scientific experiments would go to the Earth immediately. A maximum transmission rate of 46,000 bits per second, of which 6000 were needed for housekeeping, would set a limit to data-gathering from the camera and other instruments.

Real-time transmission gave the comet's dust another way of harming the mission, quite apart from gross damage to the spacecraft's structure and equipment. The transmitter power was thirty watts, so the ground station in Australia would be looking for the equivalent of a car headlamp seen at the distance of the Sun. For uninterrupted reception of data from the instruments the radio beam had to remain trained on the Earth to within one degree of the correct direction. Off-centre hits from heavy grains could set Giotto wobbling, and Lainé and others foresaw a likely interruption of the observation of the comet during the last minute of its approach to the nucleus.

'Passing within a few hundred kilometres of the comet there is a high probability that the spacecraft spin-axis attitude will be perturbed by more than one degree.' Lainé warned. 'This would occur some tens of seconds before the encounter.'

Even thirty seconds would mean a loss of communication when the spacecraft was still 2000 kilometres from the comet's nucleus. The wobble suppressors might take an hour or two to work completely, by which time the spacecraft would have left the heart of the comet far behind. The only remedy was crossed fingers.

The redesign increased the expected total mass of Giotto at launch from 840 to 950 kilograms – almost one tonne. Lainé appropriated part of the extra mass to provide sixty-nine kilograms of hydrazine fuel for the six small thrusters that controlled the spacecraft's attitude and course. It seemed far more than necessary, but Lainé mistrusted the astronomers' ability to predict Halley's movements with the required accuracy. What if the spacecraft's controllers had to make a major course correction to achieve the interception? No one foresaw the big effect that thruster power would have in Giotto's future voyagings. More striking at the time was the allocation to this fuel of more mass than all the scientific instruments put together.

Dale suffered immediate vexation. When word leaked out about a Klondike of kilograms, the scientists started a gold rush for extra mass, staking claims for fifty grams here, 500 grams there. Dale the mass miser reluctantly increased the scientific payload by three kilograms and the tussle became more feverish. One beneficiary was Susan McKenna-Lawlor. She was granted the mass to put in three energetic-particle telescopes instead of one. Lainé blamed Reinhard for the mutinous behaviour of his principal investigators, and relations between these members of Dale's team were strained.

British Aerospace was formally confirmed as prime contractor. ESTEC continued to eavesdrop on the British Aerospace computer, and Dale required any changes in the design to be settled within a month or not at all. But Link was once more master in his own house. All he had to do now was to build a spacecraft.

Whether a packet of screws came from Belgium or Switzerland was, for the European Space Agency, a matter of high politics. In a technical, budgetary and geographical juggling act, British Aerospace as prime contractor for Giotto had to assign tasks to other contractors scattered across Europe. The rule was that the industries of each member state should receive work in proportion to its contribution to the agency's budget. It was the only way of making sure that small countries did not merely hand over their taxpayers' cash to Europe's largest aerospace industries. But piecing Giotto together with bits from ten countries was a complicated game involving a chain of contractors and subcontractors.

Sources for some major items were self-evident. At Bordeaux in France the Société Européenne de Propulsion (SEP) made motors for spacecraft of just the type to throw Giotto clear of the Earth's gravity and despatch it to the comet. Indeed, the spacecraft was designed around SEP's motor called Mage – the same word for astrologer as in the artist Giotto's Adoration of the Magi. Another branch of SEP, at Vernon, supplied the electric despin motor that would keep the radio dish pointing earthwards. Alcatel-Thomson, also in France, gave Giotto its capacity to converse with the Earth by telemetry and command systems.

Giotto began to take shape beside the Bodensee in southern Germany, where Dornier constructed the airframe of aluminium alloy complete with its bumper shield. As a major contractor, Dornier also took initial responsibility for the radio antennas and the spacecraft's attitude-control system for adjusting its orientation in space.

Several components of the Dornier package originated with other contractors. While Ericsson in Sweden made the radio dish, coated with carbon fibre, Contraves in Switzerland fabricated the 'experiment platform' or instrument deck and other parts of the structure. MBB-ERNO in Germany provided electronics for the attitude control system. To find its bearings in space, Giotto required sensors for the Sun, the Earth and the stars. The Galileo company in Italy created two sets of infrared Earth sensors combined with Sun sensors. The TNO Institute of Applied Physics in the Netherlands supplied a star mapper that would either register bright stars or fix on the distant Earth as the target for Giotto's radio beam. Another Dutch company, Fokker, made the mechanical devices, called nutation dampers, which would correct the spinning of the spacecraft if dust impacts from Halley should set it wobbling.

Fokker reappeared as the contractor for the spacecraft's thermal control system, with Electronikcentralen of Denmark providing relevant electronics. The arrays of 5032 solar cells to wrap around the drum came from AEG Telefunken, as yet another piece of Giotto from Germany. FIAR in Italy devised the internal power supply system, including the batteries. Small countries had small but indispensable tasks for Giotto. In Belgium, Bell Telephone Manufacturing made electronics for controlling explosive charges that set the radio dish free to rotate independently of the spacecraft, and removed covers from some instruments. Études Techniques et Constructions Aérospatiales (also in Belgium) and Osterreichische Raumfahrt und Systemtechnik (Austria) worked on ground-support equipment.

Giotto's brain came mainly from Italy, in the form of electronic 'black boxes' supplied by Laben of Milan. The company had thirty people working for three years to perfect its contributions to the spacecraft's onboard electronics. The systems used microchips and computer logic to route commands from the Earth to their correct destinations within the spacecraft. They marshalled the data from the ten onboard experiments for transmission to the Earth. And they endowed Giotto with robot-like talents for autonomous control.

It had to be able to look after itself. In the event of any interruption of communications with the Earth, Giotto would need to know how to put itself into a safe attitude in space, that would avoid overheating or chilling and maintain an appropriate supply of solar power. It had to find the Earth again if its radio beam became misdirected. And during the hectic hours of the encounter with the comet, when events would move too fast for commands from the Earth to be effective, Giotto's own electronic systems would be fully in charge.

Many other systems in Giotto required complicated hardware and software for their control. As they were developed, they all went to Bristol for detailed testing of their performance. Then the contractors' engineers took their software to a simulator at ESTEC where they sat down with agency engineers to search exhaustively for possible bugs in the programs. Every time they found one, the testers put a champagne cork on top of a board in the room. Eventually there were many corks, each one representing a hidden flaw in the software that might have disabled Giotto, but didn't.

The paperwork describing the spacecraft and its subsystems ran to 500 volumes, representing a huge range of engineering skills, from materials science to computer programs. David Link and his team at British Aerospace had to understand all the technicalities well enough to coordinate the manufacturing, assemble a fully functioning spacecraft, know all its idiosyncrasies, and teach the mission controllers how to fly it.

They built various versions of Giotto. A mock-up helped to show how all the pieces would fit together and how they could best be assembled. One version of the spacecraft completed by Dornier underwent severe tests to check the integrity of the mechanical structure. Another would be used for verifying the electrical functions of the spacecraft at Bristol. There was also a set of spare units. The 'real' Giotto, treated with reverence, was the flight model that would travel to Halley's Comet. It too had to undergo searching tests in other European centres. The Bristol team devised trolleys, packaging and other equipment, for the safe transport and handling of Giotto on its terrestrial wanderings before it reached the launch pad.

CHAPTER FOUR

RACING A COMET

FOURTEEN months after its closest approach to the Sun in 1910, Halley's Comet was last photographed disappearing towards the outer Solar System in June 1911. Astronomers hoped to find the comet at a much greater distance when it flew in for its 1986 appointment with the Sun. In modern ground-based telescopes, electronic light detectors including the supersensitive charge-coupled devices, or CCDs, were replacing long-exposure photography. But the best instruments in the world would be useless unless pointed to the right place in the night sky.

Donald Yeomans at NASA's Jet Propulsion Laboratory predicted the comet's position on its orbit, and in the Earth's sky, for any time when astronomers chose to look. Because he had checked all peculiarities in Halley's motions, right back to classical times, his computations were likely to be far more accurate than those available at the 1910 apparition, when the comet ran a full ten days behind the predictions.

The most determined Halley hunter was Mike Belton of Tucson, Arizona. He was already looking for it intermittently, with a large telescope on Kitt Peak, Arizona, while he served on the US-European team planning the International Comet Mission in the late 1970s. The comet was then still farther away than the planet Uranus. Belton's hopes intensified in the early 1980s. Estimates of the brightness of a cold comet lit by a distant Sun suggested it should come within range of his telescope somewhere between the orbits of Uranus and Saturn. In mid-October 1982 Belton briefly interrupted his search. Knowing the comet's expected route like the back of his hand, he could see it would pass nearly in front of a star so bright as to dazzle his detectors.

Another of the Earth's big eyes, on Palomar mountain in California, was searching for faint galaxies, still the more serious business of mainstream astronomy. David Jewitt and Edward Danielson managed to borrow the telescope for an hour or two a night, when the galaxy people would let them, together with one of the precious arrays of CCDs. Working less systematically than Belton, they pointed the telescope in the expected direction of Halley. The bright star in the field of view came as an unpleasant surprise. In the hope of masking the star and seeing the comet, the astronomers used razor blades to cover all but a narrow strip of the CCDs at the focus of the telescope.

It was 16 October 1982. As Jewitt and Danielson exposed the detector with its improvised mask at the focus of the telescope, the Palomar mirror inched around like the hour hand of a clock, to correct for the spin of the Earth. It was not programmed to move with the stars, in its usual fashion, but to follow Yeomans' predicted track for the comet. When they processed their electronic data, the astronomers found streaks made by faint stars and one sharper image of an object on the predicted track.

Seen at an astonishing distance of more than 1.5 billion kilometres, Halley's Comet was coming back. Belton was beaten at the post, but Yeomans at JPL had every reason for congratulating himself. He remarked, 'If you can predict the comet's motion well for 1986 and again in 164 BC, you've got to be doing something right.'

Yeomans' own role in the 1986 event then began in earnest. During the forty-one months remaining until the arrival of the international fleet of spacecraft at Halley, he would use many telescope observations of the comet to refine his predictions. That would give the mission controllers the best possible chance of intercepting the comet in the manner they intended.

News of the comet's reappearance electrified the Giotto team. Their mission had depended on a mathematical calculation, and jokers could cause at least momentary qualms by asking, 'What will you do with all this junk if the comet doesn't show up. What if it's fallen in a black hole out there?' Now the project leaders had a slogan: 'Halley's Comet is on schedule. Let's ensure Giotto is too.'

A master chart in the office of the project's schedule controller at Noordwijk governed everyone's life. In intricate, interlocking detail it showed what had to be ready when, if Giotto was to rise from the launch pad in time to keep its appointment with the comet. Any slippage would mean missing the target by three-quarters of a century.

David Dale and his management team turned Nature's timetable to their advantage. Spacecraft development was plagued with delays that could run to months, years or even decades. The high costs and risks of any mission, and the large numbers of talented specialists involved, gave endless scope for chats about improvements and precautions – for 'debating societies' as Dale called them. When suggestions and arguments arose during Giotto's development, Dale would listen politely and then cut the conversation short. He wanted an immediate decision that preferred the safe and adequate solution to any fancy variations.

Giotto's experimenters, courteous anarchists by temperament, were the trickiest to manage. They knew that the only purpose of the mission was to carry to Halley the instruments supplied by them and paid for by their national funding agencies. Rüdeger Reinhard had a deeper understanding of the scientists' ambitions and technical difficulties than any other member of the management team. He was the perpetual go-between, using his knowledge and ingenuity to judge when to argue in support of the scientists and when to crack the managerial whip.

Reinhard gathered the principal investigators together for frequent meetings as a science working team. This kept everyone's minds focused, not just on the mechanics of completing the spacecraft, but on the adventure that awaited all of them in the few hours when their instruments came face to face with Halley. The group cultivated a fine team spirit but it created more than a 'debating society'. It threatened to become a caucus that could question Dale's decisions. As a prophylactic against time-wasting, Reinhard decorated the minutes of the science working team with a map of the Solar System showing the position of the comet on the date of the meeting.

Dale demanded monthly reports on the mass, power requirements and state of readiness of every instrument under development. As nobody wanted to be left behind when Giotto flew, Dale found the scientists easier to deal with singly. 'Yes, I mean it, you must save twenty grams.' 'No Susan, you can't add on a dust experiment from Chicago.' When Reinhard seemed too gentle with them, John Credland could always play Mr Nasty to Dale's Mr Nice, and unleash his Yorkshire tongue. As a leading member of the team of some seventeen engineers working for Dale, Credland was responsible for the integration of the experiments in the spacecraft. A dangerous ambiguity remained about who was ultimately accountable for the state of Giotto's instruments, the scientists or the project management.

Then there was Fritz Neubauer fussing around his fellow-scientists' instruments like a supernumerary project engineer. This bespectacled physicist from Cologne, with his jaw cradled in a sleek, dark beard, was in charge of Giotto's magnetic sensors. As supplied by co-investigators from NASA's Goddard Space Flight Center, the sensors were small high-frequency transformers, in which the frequency harmonics of the output changed in the presence of external magnetic fields. But Neubauer's hopes of measuring the subtle magnetism of the solar wind, and seeing how it changed in interactions with Halley, were threatened by interference from the spacecraft's experiments, power supplies and electric motors.

In most spacecraft, magnetometers went out on the ends of long booms, to escape the stray magnetic fields of other equipment. With no booms allowed on Giotto because of the dust hazard, Neubauer's instruments had to go on the tripod that surrounded the radio dish. David Dale left it to Neubauer to make sure that the spacecraft was magnetically clean enough for his observations.

This meant counselling engineers and experimenters about the design of their equipment, and then mapping the magnetism of the finished components item by item. Neubauer's team used a magnetic coil facility at the Technical University of Braunschweig. When that was not possible, they took a mobile system to the spacecraft during its assembly at Bristol. They planned to follow Giotto around as it went on to tests in France and at the launch site. Neubauer also arranged for the finished flight model of Giotto to visit Germany before the launch, where the IABG (Industrieanlagen-Betriebsgesellschaft) had a facility suitable for magnetic tests on the whole spacecraft.

He knew he would never achieve perfect magnetic cleanliness, but he could at least minimize fluctuations in the magnetism of the various components. Overall, for most of the equipment in Giotto, the results were almost as good as if the magnetometers were on long booms. Two troublesome items remained. The despin motor of the radio dish and the motor that aimed Keller's camera both emitted strongly fluctuating magnetic fields. But these were at least predictable, and Neubauer expected that he would be able to correct his observations by comparing the measurements of two magnetometers at different positions on Giotto's tripod.

Those working on other space projects, sometimes more complex and expensive than the comet mission, envied the priority accorded to Dale's team throughout the European Space Agency. The name Giotto was like a stamp saying 'Attention – urgent'. So when delays became apparent in the new and more powerful generations of Ariane launching rockets, sponsored by the agency, a change of plan gave the Halley mission a launcher all to itself.

Since the earliest days when Giotto was supposed to have a fraternal twin,

Geos-3, the idea of saving a lot of money by a dual launch had always been part of the scheme. 'Someone else's satellite' would ride into space with Giotto on an Ariane 2 launcher. The French-led Ariane development was going very well on the whole, but the rocket engineers were not spared the occasional failure that kept them and their clients living on a knife-edge, and left the launchers grounded for months on end while the engineers traced and cured the fault.

When the Ariane 2 launcher was plainly running late, and new generations were overtaking it in the development workshops, Giotto was switched to an Ariane 3. A poser for Arianespace, the company that marketed the launches, was to find a customer willing to be paired with a scientific mission with a narrow range of possible launch dates. No one with an expensive satellite to put into orbit wanted to be pinned down so precisely to early July 1985.

The solution came with the launch in May 1983 of Europe's X-ray astronomy satellite called Exosat. An Ariane 1 rocket had been booked for a back-up launch, but an American launcher sent Exosat safely into orbit at the first attempt. There was an Ariane 1 to spare, which Giotto could have.

'Will it cost me any more?' was David Dale's first question.

'No,' Arianespace said, 'you get it at the agreed price.'

The rocket was far too powerful for the job. In fact it could send Giotto direct to Halley without any help from the spacecraft's onboard motor. Why not leave the motor out? Dale quickly suppressed any debate about that option. Playing safe as always, he rejected so violent a change to the original mission plan halfway through the development of Giotto, when all design of flight hardware had just been agreed and frozen. 'To take the motor out would affect the overall structure, the bumper shield, the payload decks, and so on.' Dale said. 'An empty motor shell would upset the balance during launch. And we can't leave the propellant in there unfired and fly it to the comet.'

Even though the onboard motor would remain, the solitary launch by Ariane 1 still demanded a heavier spacecraft. Arianespace devised a manoeuvre in which the rocket would dip down towards the Earth for a while in mid-launch, to shed surplus energy, before resuming its climb to orbit. Despite that ingenuity Giotto would have to be made 10 kilograms heavier, 960 instead of 950 kilograms, if necessary by adding lead weights. Scientists who had suffered misery to save a few grams found this grotesque.

Dale was unrepentant and went on husbanding mass jealously. He required reserves of mass to enable the team to solve critical engineering problems with the thermal control system and the despin motor. He had to be able to dole out

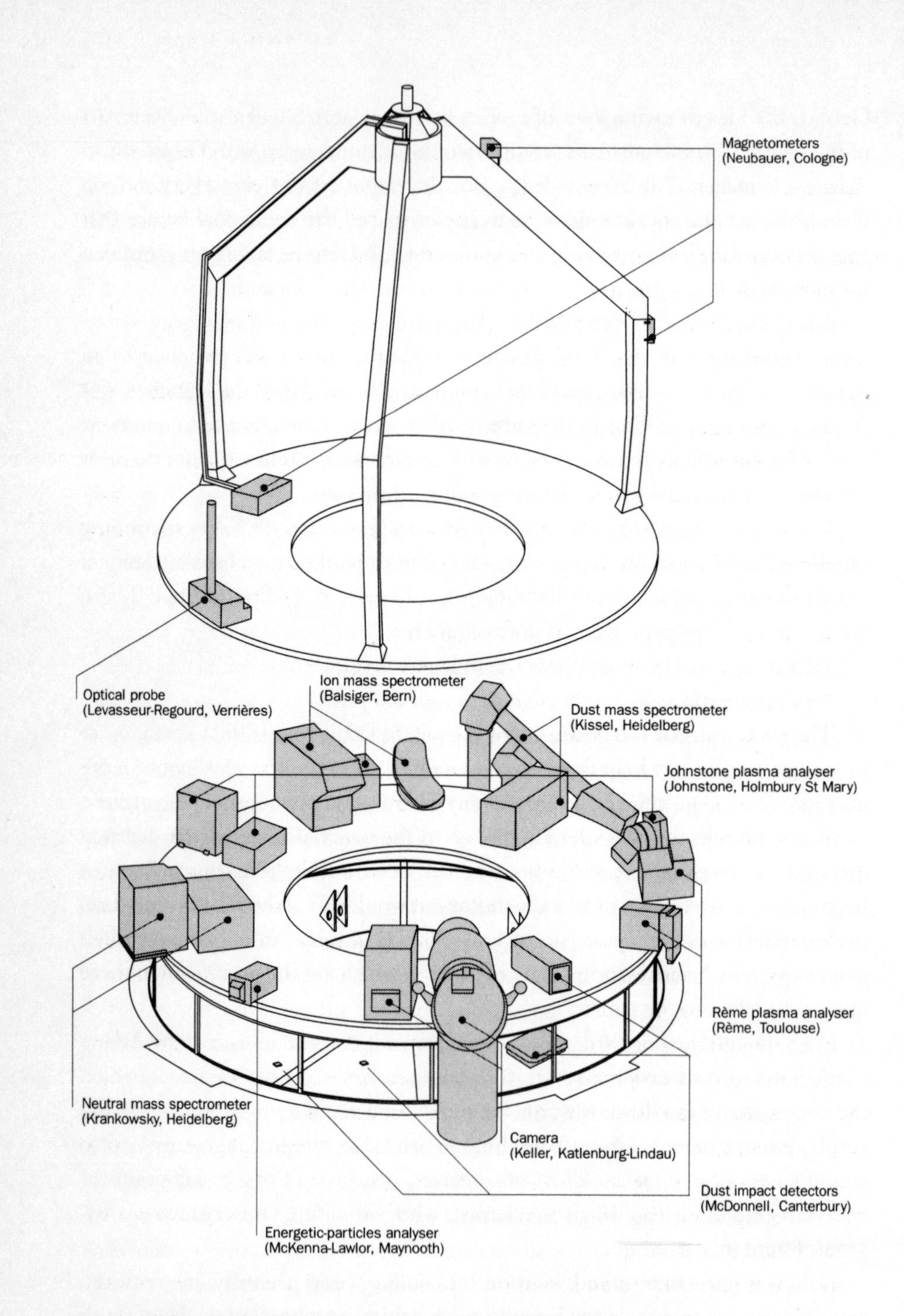

The scientific instruments packed behind Giotto's bumper shield, at the bottom of the spacecraft. The names and addresses are those of the principal investigators.

teaspoonfuls of extra mass to Keller and other scientists who proved they needed it. Two years ahead of the launch Dale was already looking at a total mass of 930 kilograms, and who knew what other problems might arise? In any case he might want twenty kilograms of lead weights simply to balance the completed spacecraft so that it would spin true without wobbling. As for the scientists' gripes, they were as predictable as the sunrise.

On the heights overlooking the Aare river, where it looped through Bern, Hans Balsiger was busy with Giotto's ion mass spectrometer. Switzerland's federal capital was famous for its bears and its chocolate, but scientists worldwide knew its university for a 20th-Century tradition in measuring small traces of elements in materials. This work relied on a series of mass spectrometers, which counted individual atoms. The physics institute, where Balsiger was working, had applied these instruments in studies that ranged from climate changes on the Earth to the composition of lunar rocks brought back by the US astronauts. Here Johannes Geiss analysed the heavy atoms in the solar wind, using the metallic 'Swiss flag' deployed on the Moon.

Balsiger began his own research career at Bern by measuring abundances of titanium and lithium in meteorites, and he made a start in space research in the early 1970s by flying a mass spectrometer into the upper atmosphere on rockets. This looked for extraterrestrial materials deposited there by meteorites and comet dust. The European Space Agency's Geos mission of the mid-1970s, to which Bern contributed an ion mass spectrometer, introduced Balsiger and his colleagues to the exacting task of making a lightweight satellite instrument that would survive the violence of the launch and function perfectly in the vacuum of space. They built at Bern a large vacuum chamber, unique in Europe, in which instruments could be subjected to encounters with charged particles, simulating the conditions they would meet during the mission.

A mass spectrometer uses magnetic and electric fields to separate electrically charged atoms or molecules, according to their different masses. Giotto was to contain several mass spectrometers for investigating Halley's chemical composition. They were distinguished chiefly by the sources of the materials to be analysed. Neutral atoms and molecules emerging from Halley were the concern of the group led by Dieter Krankowsky of Heidelberg, with Peter Eberhardt from Bern contributing a substantial part of the hardware. Analysis of the contents of dust particles was also masterminded at Heidelberg, by Jochen Kissel.

Balsiger's interest was in the ions, that is to say, atoms and molecules that would arrive at the spacecraft already possessing an electric charge. The small instrument added by Axel Korth of Katlenburg-Lindau to Henri Rème's plasma analyser was to measure relatively heavy ions from the comet. Balsiger's ion mass spectrometer would cope with a range from hydrogen nuclei, the lightest and commonest, up to ions with the mass of iron atoms. It was to consist of two different sensors, called HERS and HIS.

Marcia Neugebauer of NASA's Jet Propulsion Laboratory provided the concept of HERS. This was a mass spectrometer adapted to the outer regions of the comet's head, where 'hot' ions of solar wind would be mixed with ions picked up from the comet. For 'cool' ions of cometary origin, best detected near the nucleus, the HIS mass spectrometer was designed by Helmut Rosenbauer of Katlenburg-Lindau. At first these were to be in the same box, but they finished up as separate items.

Balsiger himself developed the magnets for the experiment, but as principal investigator he was also responsible for coordinating all the work on both sensors, and for their calibration and testing. He found it a nightmarish experience. To suit the special environment at Halley the experimenters had to produce designs that had not been tried in space before. That would have been exacting without the irrevocable timetable of the mission. When delays with vital components put his experiment behind schedule, Balsiger was under such stress that his health suffered. Geiss took over the management of the experiment for three weeks, to give Balsiger a rest from Giotto.

Outsiders might imagine that the day-to-day technological problems with Giotto's instruments drove from the scientists' minds the grand objectives of the mission. On the contrary, what kept them going was the conviction that well-made experiments would achieve discoveries in cosmic chemistry, and answer specifiable questions about the nature and roles of the comets. It moved them to work to the limits of their endurance.

At the mission's most westerly outpost, in the Irish town of Maynooth, Susan McKenna-Lawlor was having the time of her life. To make her energetic-particles analyser she was creating an Irish tradition in space research and technology virtually from scratch. Hers was the smallest and neatest of Giotto's experiments, but from the outset she was aware, from her experience with NASA, of the special facilities and fabrication skills needed for making any space-qualified experiment.

Through her association with the Max-Planck-Institut at Katlenburg-Lindau, McKenna-Lawlor had primary help with the assembly of the first version of the instrument, for engineering tests. With backing from Ireland's National Board for Science and Technology she set up her own clean room and test facilities at Maynooth, and gradually built up a home-grown team of design engineers.

She finished the flight model of the instrument on time, two years before the launch, as required in the mission plan to give British Aerospace time to integrate it in the spacecraft for subsequent tests. Not everyone was so punctual. But even then McKenna-Lawlor did not sit back. She felt personally responsible for Ireland's first space experiment and insisted on observing all the relevant tests herself, even if that meant staying up all night.

For Uwe Keller, racing a comet meant indulging in his habit of driving too fast. After a visit to Liège, where his Belgian co-investigators had a facility for testing Giotto's camera and its components in a simulated space environment, he was approaching Cologne along the autobahn at his customary 200-plus kilometres per hour. A truck pulled out in front of him, and seconds later Keller's Porsche was shooting backwards down an embankment planted with trees. The car threaded its own way through the trees, came to rest, and caught fire.

Keller scrambled out and walked away unhurt, but not before he had snatched the camera test results from the burning vehicle. His reaction to the accident was that he might have avoided trouble with a little more acceleration. He bought a faster car.

At his laboratory at Katlenburg-Lindau, Keller wrestled with problems that would make a less stubborn person give up, especially if he never wished to be principal investigator for Giotto's camera in the first place. The enterprise was risky. He was working with CCDs, the charge-coupled devices so innovative that Keller could not obtain them from Europe but had to find them at Texas Instruments. He and his team were piecing together mechanical, optical and electronic systems into a television camera of a kind never made before. On top of all that, the whole damned thing had to be smart.

Keller had a vivid picture in his mind of Giotto spinning into the dust clouds of Halley's Comet and dashing past the long-sought nucleus at 68.4 kilometres per second. It was a speed equivalent to travelling from the Earth to the Moon in ninety minutes. His earlier simile, of trying to see a star through a telescope while riding on a carousel, conveyed the problem of working from a spinning spacecraft, but it greatly understated the task. When a journalist asked why the camera had as much computing power as the rest of Giotto put together, Keller offered another analogy.

'An aircraft is passing you at supersonic speed and you're trying to take a series of photos of the pilot in the cockpit. And if you once lose sight of him you won't have time to find him again.'

The camera's operation had to be self-sustaining. From the moment the power was switched on by a command from the ground, when still a million kilometres from the comet, the camera would first go into a search and acquisition procedure. It would scan the sky by rocking a mirror, looking for the comet's nucleus, somewhere ahead of the spacecraft. Its instructions were to search for the brightest spot in its vicinity, though this was not a decision taken lightly.

Keller asked the members of his team to sketch their ideas about what the nucleus might look like. One danger was that a dust cloud in the heart of the comet might be brighter than the nucleus itself. But any more subtle instruction, such as 'Look for the region of greatest contrast', would be very risky. Besides being more difficult to define and compute, it would involve a prejudice about the appearance of the nucleus that could turn out to be completely wrong.

When it found the presumed nucleus, the camera would start taking pictures, by exposing its CCDs at exactly the right moment during each rotation of the spacecraft. To prevent smearing of the pictures, the exposure would be only fourteen microseconds at the time of the closest approach to the nucleus. But the camera would lose sight of the nucleus unless it kept adjusting itself to point the right way from shot to shot. Therefore its electronic control system had to compute the relative motion of the nucleus, based on its own estimate of the distance and time to Giotto's moment of closest approach. Then a motor had to swivel the camera at the right rate to keep it trained on the comet.

While all the reckoning was going on, in the camera's cerebellum, the rotation of the spacecraft would keep changing the apparent position of the comet in relation to the camera far faster than the changes caused by Giotto's forward motion. A possible imbalance in the spacecraft aggravated the computational task. Despite their precious lead weights, the engineers could not promise to eliminate the wobble entirely. 'It might be as much as four-tenths of a degree,' they told Keller. That was equal to the camera's whole field of view, and it could throw the nucleus right out of the picture. A raft of extra software went into detecting the wobble and correcting for it. Finally, Keller's team had to test all the computing programs rigorously, before burning them into the hardware of programmable microchips.

Keller thought he had quite enough to worry about, without endless grumbles from the project management. During a visit to Katlenburg-Lindau, David Dale complained that the mass of the camera and its control system was creeping well beyond the original limit.

'You'll have to use titanium screws,' Dale said. 'That will save fifty grams.'

'They would cost 50,000 marks!' Keller protested. Dale was unmoved, but he stopped pressing the point when Keller told him that to have titanium screws manufactured would take six months. Keller thought the argument showed that Dale made too little allowance for the importance of the camera and its exceptional complexity. The management was supposed to help everyone meet the deadlines, in a spirit of teamwork, yet seemed to compound Keller's difficulties by always insisting that the camera was only one experiment among ten, and that all principal investigators were equal.

Had Keller been a mind-reader, he would have seen that his camera was in truth more equal than others. Skirmishing about screws and grams was the only way the project manager could try to push Keller along. Dale dreaded the blame that would fall on him if he delivered Giotto to Halley and it were blind.

Dale respected Keller as a scientist. He had met bright individuals working on the technicalities of the camera, and was impressed by its complexity. But in the institute at Katlenburg-Lindau there were none of the careful schedules that decorated the project office in Noordwijk. For two years the camera team looked to Dale like a debating society that never finished its business, or made the systems-level decisions that were urgently necessary.

'Where's your plan?' Dale kept asking Keller, meaning straightforward manufacturing schedules.

'Everything will be OK,' Keller said, thinking of what would be written on the programmed microchips.

'Where's the evidence?'

'Oh you worry too much, Dave.'

Keller spoke that way only to get the project management off his back. As the principal investigator, he had more reason than anyone else to worry about the slow progress with the camera. His engineers were working hard and effectively, but there were not enough of them.

The aeronomy laboratory at Katlenburg-Lindau could well have been renamed the Max-Planck-Institut für Halleyschen Kometen because it was involved to varying degrees in no fewer than nine encounter experiments – five in Giotto and four in the Soviet Vegas. Amid in-house competition for technicians, facilities and funds, Keller had less help from his own institute than he had been led to expect.

Suddenly time was running out. By July 1983, when all of Giotto's experiments were supposed to be ready, most of the pieces of the camera were available but it was nothing like a fully-functioning system. By later in the year, Dale and his team were becoming convinced that the camera could not possibly be finished by July 1984, and even having it ready in time for the launch itself, in July 1985, looked doubtful. Was Jacques Blamont right all along, when he said that 'amateurs' in Europe could not do the camera?

Somebody had to take drastic action, about the support given to Keller, and about money too, because every slow step forward added to the expected bill. Funding the camera was primarily a German responsibility.

'What we need is a German spotlight on Keller,' Dale said to Roger Bonnet. The French space scientist was Dale's boss since May 1983, when he succeeded Ernst Trendelenburg as director of science in the Paris headquarters of the European Space Agency. His mop of hair was now a distinguished white.

Bonnet himself was listed as a co-investigator with Keller, and his former institute at Verrières-le-Buisson was supplying the camera's optical system. As both researcher and administrator for the same operation he found himself *pris entre le marteau et l'enclume* – caught between the hammer and the anvil – when he attended a meeting with Keller and all his co-investigators at the Ringberg Schloss near Munich in November 1983.

Keller had a rough ride even among his friends who knew about the shortage of engineers at Katlenburg-Lindau. He was already looking for remedies himself when he took a nocturnal walk in the snow with Alan Delamere, the camera's inventor, who was over from the US. Keller asked if Delamere would be willing to come to Germany and help to get the camera finished. He said Yes, in principle he would, if the move could be squared with his company, Ball Aerospace.

What Bonnet heard at the Ringberg Schloss persuaded him that Dale's misgivings about the camera were justified, so he switched on the brightest German spotlight he knew. He phoned the president of the Max-Planck-Gesellschaft, proprietor of the nationwide network of Max-Planck research institutes of which Keller's was one. The president was due to become the next director general of the European Space Agency.

'Before you come to Paris,' Bonnet suggested, 'hadn't you better resolve this crisis over Giotto's camera?'

While he protested to Bonnet that each Max-Planck institute was supposed to manage its own affairs, the president acted swiftly behind the scenes. Warning the Bonn government that Germany's reputation in technology was at stake, he descended on Katlenburg-Lindau to investigate the status and funding of the camera project. A new acting-director was going to take over the institute on 1 January 1984, which made policy changes easier. In the New Year the Giotto camera was given priority over all other activity in the institute. Lack of money, Keller was told, was not to delay the work. Extra funding from the Max-Planck-Gesellschaft paid for Delamere to come from the US to manage the camera's completion.

With these reforms in prospect, Bonnet, Keller, Dale and Reinhard could all set off to a meeting in Japan, just before Christmas 1983, in a more cheerful mood.

That Halley's Comet should bring officials of opposing superpowers together in a spirit of peaceful scientific collaboration was nothing to take for granted at the end of 1983. The planet Earth was in a worse state than usual. Wars raged between Iran and Iraq, between the countless factions in Lebanon, and between the Soviet army and Afghan rebels. In the summer the Soviet air force shot down a Korean jumbo-jet full of passengers. The thermonuclear arms race of the US and the Soviet Union was rushing on, with the US President calling the Soviet Union the Evil Empire and Soviet negotiators walking out of disarmament talks in Geneva in protest against the US 'star wars' programme for strategic defence systems in space. East-West relations were at their coldest for ten years.

In those unpromising days, at a meeting at the Japanese space base at Kagoshima on 18-19 December 1983, representatives of the world's leading space agencies decided on highly practical cooperation. Its main purpose was to guide the European spacecraft Giotto as accurately as possible along its chosen track close to Halley's nucleus. This was Pathfinder.

Europe's representatives had sketched the idea of Pathfinder at the first meeting of the Inter-Agency Consultative Group, held at Padua in Italy in 1981. The European Space Agency, NASA, the Soviet-led Intercosmos agency, and Japan's Institute of Space and Astronautical Science were members of the group. As the Soviet Vega spacecraft were due at Halley a few days before Giotto, they could help to pinpoint the position of the nucleus and so improve Giotto's aim. American assistance would be imperative. After two years of technical study and diplomacy, the third inter-agency meeting at Kagoshima decided that Pathfinder should happen.

A Hungarian scientist with the Intercosmos party commented: 'Right now this is the only major scientific cooperation in space. We show a very positive example in these difficult times.'

What seemed to outsiders like a political miracle was fairly easy to accomplish at the level of individual officials, scientists and engineers. They knew one another. Even during the Cold War's chilliest phases, space scientists met frequently at international symposia, and the agencies had previously cooperated with one another in projects large and small. Their common interest in Halley's Comet, and in what it might reveal about the world's origins, was untainted by ideology.

Giotto's project scientist Rüdeger Reinhard, who doubled as secretary of the Inter-Agency Consultative Group, was chairman of its working group on the Halley environment. He and Uwe Keller, together with old friends from the US-European comet team, shared with their opposite numbers from the Soviet Union and Japan all their studies of the dust hazard to spacecraft venturing near Halley. Another working group brought together scientists whose spacecraft instruments would be studying the solar wind near Halley, so that they could coordinate their simultaneous observations.

Pathfinder evolved in a third group concerned with spacecraft navigation and mission optimization. The head of NASA's Deep Space Network promised that his ground stations would work in combination, using the radio astronomers' technique called very-long-baseline interferometry to track the Soviet Vega spacecraft with extreme precision during their encounter with Halley. Images and other data from the Vegas would show the distance and direction of the comet's nucleus from about 10,000 kilometres away. Roald Sagdeev, the Soviet Union's affable director of space research, promised that Intercosmos would pass this information to the European Space Agency quickly enough for steering Giotto.

Putting the US and Soviet data together, the European Space Operations Centre at Darmstadt could reduce the doubts about the position and track of the nucleus, for the final course correction in Giotto's approach to Halley. The best efforts of ground-based astronomers to observe the comet, and of Donald Yeomans to compute its track, were likely to leave an uncertainty of a few hundred kilometres. Pathfinder should reduce it to some tens of kilometres.

David Dale found himself cast in a diplomatic role as chairman of the steering committee that would oversee Pathfinder. Despite Sagdeev's enthusiasm the Soviet bureaucracy stuck to its old doctrine that no visa should be issued until the train was ready to leave. The formal agreement about Pathfinder was signed less than two months before the Vegas and Giotto arrived at Halley. Experts from ESOC nevertheless set to work with their opposite numbers from NASA and Intercosmos to coordinate the technical arrangements.

Reinhard authorized a 'debating society' in his science working team, by inviting Giotto's experimenters to assume that Pathfinder would give the expected precision and to consider just how close to the nucleus they would like the spacecraft to go. The nominal distance had always been 500 kilometres. Keller said that it should not be closer, or the camera could not swivel fast enough as it went past the nucleus. He would prefer a miss distance of 1000 kilometres. All of the other scientists favoured going as close to the nucleus as possible, but they were divided on what 'possible' meant. For some, Giotto should survive the encounter, while the others would not mind if the Halley dust-storm destroyed it. The group deferred the decision until the last minute, when they would know the accuracy of the Pathfinder results.

On 22 December 1983, a five-year-old spacecraft called ISEE-3 swung around the Moon and changed its name to ICE, for International Cometary Explorer. It was the spacecraft with which NASA put itself back into comet raiding after the fiasco about Halley. This lunar manoeuvre was the last of five, over a period of eighteen months. The first sent ISEE-3 out 1.5 million kilometres into the Earth's magnetic tail. On its return, further complicated manoeuvres, all figured out by Robert Farquhar of the Goddard Space Flight Center, altered the spacecraft's orbit and speed by stages until it finished up on course to intercept Comet Giacobini-Zinner. ICE was expected to fly by that comet six months before the international fleet's arrival at Halley. Finally, ICE would become an outrider of the Halley fleet, passing 28 million kilometres from the big comet.

In Bristol as in Noordwijk, the race with Halley's Comet gave Giotto a special status. David Link lead a team of total enthusiasts at British Aerospace. They were creating the vehicle for a great venture of the human spirit. It was to plunge deeper into the dust clouds of the comet than the Soviets dared to go, and no engineer could wish for a more exciting challenge, and no craftsman a better destination for his handiwork. The hardware had to be fault-free, the software bug-free.

To the multicultural European venture, the British brought their capacity for humour and poetry. A publication called *GOB* was the unofficial mouthpiece of the Giotto project at British Aerospace. Modelled on the satirical magazine *Private Eye*, it was full of scurrilities about managers and other misfortunes. And Rod Jenkins, the chief project engineer at Bristol, collected the poetry. He treasured verses by U.A. Fanthorpe, who depicted Halley's Comet as an athletic paramour of the Earth:

This time my lover sends me dancing-partners
Whom I shall shatter in my dusty unveiling.

In what little spare time he had, Jenkins immersed himself in the astronomy and folklore of the comet. He familiarized himself with the story of Newton and Halley and found a local church with a wind vane modelled on Halley's Comet at its first predicted reappearance. He matched this with a contemporaneous report on the comet that he unearthed in a Bristol newspaper of 1759: 'The expected comet now appears every clear evening, till ten or eleven o'clock, to the west of the south, under the constellation of Hydra.'

Jenkins found words of his own to express what many of his colleagues across Europe felt about the mission but were too shy to say. Professional pride in their technical tasks, and fear of widespread reproaches in the event of failure, were insufficient to explain the zeal with which so many individuals in so many countries worked long hours to make Giotto a success.

'It is an opportunity for mere mortals to associate with the gods and giants of the past,' Jenkins declared. Such conviction was needed in ample measure when the project hit its inevitable snags. Paint, for example, perplexed the engineers for more than two years.

The bumper shield and other parts of Giotto were to be coloured white to reject the Sun's heat. The white paint also had to be electrically conducting to prevent dangerous voltages building up when the spacecraft was enveloped in electrified gas during the encounter. It was a difficult combination of properties to achieve, especially as some ingredients were banned because they would contaminate Giotto's chemical instruments.

After a British laboratory had failed to produce the right paint by early 1983, the Centre National d'Études Spatiales in France provided paint that it had developed independently. Extensive tests in France and at Noordwijk preceded its arrival, but when British Aerospace tried it, the paint flaked off like falling snow. The problem was solved only by close attention to the paint's formulation and to the precise hand movements of the painter when applying it.

If the camera crisis was the chief headache on the scientific side of the project, the despin motor caused the biggest worries for the spacecraft engineers. The mission literally revolved around this motor beneath the radio dish. Giotto could send home no information about Halley unless the dish stayed trained on the Earth while the rest of the structure rotated around it. The SEP company at Vernon in France was the brave contractor for the despin mechanism, which was of a novel design.

Ordinary lubricants could not work in the vacuum of space, and the dry bearings for the radio dish had to pass searching examinations at a friction laboratory in England. But when the first, structural model of Giotto underwent fierce tests in Germany, in which vibrations, intense noise and centrifuging simulated the stresses to be experienced during launch, the despin mechanism turned out to be vulnerable. The designers had to look again at the 'offloading' system that protected the bearings during the launch. A worse shock came at Bristol late in 1983, in the version of Giotto built for electrical tests. The despin motor kept sticking. British Aerospace, ESTEC and the contractor began urgent consultations about fixing it.

Then, much sooner than anyone expected, Giotto found itself in France. The completed flight model was scheduled to leave Bristol in the summer of 1984 for a long programme of final tests in France and Germany. At the beginning of the year, though, Link was uneasy about industrial relations within British Aerospace as a whole. Disagreements between the trade unions and management were distracting even his own resolute workforce. A strike could threaten Giotto's schedule fatally.

Although he knew the move would be a big upheaval, Link decided to take the flight model out of the Bristol factory half a year ahead of schedule. This turned out to be a crucial management decision. Giotto was whisked off to the Intespace test facility, in the French government's space complex at Toulouse, just before the feared strike closed the Bristol factory in February.

Craftsmen from Bristol went to Toulouse too, to continue with their work. Halley was still hurrying towards the Sun, unaware of the grievances that prompted the strike. The trade unions that organized it instructed their members in Toulouse to return to the United Kingdom, or else give their wages to charity. The dedication to Giotto was such that nearly everyone chose the second option and the assembly of the spacecraft continued.

They were anxious days for Terry Harris, the British Aerospace manufacturing supervisor, who had to hold the workforce together by tactful leadership, not only within the team but among a colony of Bristolian families too. All over Europe, those busy with Giotto needed the unpublicized support of their partners at home, as they worked ridiculous hours, absented themselves abroad and skipped their holidays. The company arranged for the technicians' families to join them in Toulouse, but the wives argued among themselves about whether their husbands should go on working during the strike.

In an action without precedent in spacecraft engineering, the families were taken to see the flight model of the spacecraft at Intespace. Small children, garbed in clean-room coats, hats and overshoes many sizes too large for them, shuffled into the holy of holies and viewed the famous Giotto in all its gleaming intricacy. This visit by the families helped to clear the atmosphere in the workforce. Back in the United Kingdom the strike was over after eight weeks, but the exile of the British Aerospace team in Toulouse lasted for a further year of assembly and testing.

A blanket was missing from the spacecraft's insulation system, just when it was due for thermal tests to see how it would cope with extremes of heat and cold in space. John Credland, who represented ESTEC's project management at Intespace, established that the material was still in the Netherlands. A secretary at Noordwijk was willing to fly with it to Toulouse, but clearing the expensive metallic foil through French customs could take some days. With the whimsy of a true Giottician, she dressed up as an artist and was ready, if challenged, to say she used it for her collages. That would have been nearer to the truth than she knew, because she found herself working with scissors at three in the morning, as a scratch squad put the thermal blanket in place before the experts arrived to start the test.

A fire at Noordwijk in November 1984 disrupted the work of David Dale's team. The Giotto project offices were on the floor above an acoustic test chamber, and material in the chamber walls caught fire during welding work. Firemen put out the blaze, but not before six offices were badly damaged. Charred papers and barbecued computer terminals told the Giotto team at a glance that much work had gone up in smoke.

The project's route map, the master chart in the schedule controller's office, was scorched black. No one had a copy, but the original was just legible when held against the light. As soon as the team was installed in fresh offices, the schedule controller and his helpers were straining their eyes to transcribe the schedule on a drawing board. After dark, the work continued with the aid of a spotlight.

At the European Space Agency's Paris headquarters, Roger Bonnet was promoting a long-term programme of research missions that would attract increased funding from the agency's member states. Called Horizon 2000, it was based on ideas solicited from far and wide among Europe's scientists. Meeting in Venice, a survey committee made a balanced selection of four major projects to be pursued on a long-term basis as cornerstones of the agency's science programme. Two were astronomical missions, for advanced studies of the Universe by X rays and submillimetre radio waves. Another was a dual mission called Soho and Cluster for observing the Sun and the solar wind.

The fourth cornerstone was an ambitious comet mission, intended to gather samples from the nucleus of a comet and return them to the Earth for analysis. It would probe deeper into those questions of cosmic origins, to which Giotto might give only tentative answers. This projected mission came to be called Rosetta. Europe's comet scientists hoped it meant that the European Space Agency would not lose its interest in comets when the Halley flyby was over.

CHAPTER FIVE

GIOTTO JOINS THE FLEET

In the midwinter of 1984-85, three spacecraft of the international Halley fleet departed on their missions. On 15 December a Proton rocket rose from the Soviet cosmoport in Kazakhstan to send Vega-1 on its way, first to Venus and then onwards to Halley's Comet. The name Vega was a contraction of the Russian Venera (Venus) and Gallei (Halley). Six days later Vega-2, which was identical, set off on the same magnificent mission, more complicated than Giotto's.

Each Vega spacecraft weighed 4.5 tonnes at launch, compared with slightly less than a tonne for the European spacecraft. It would shed an instrumented balloon and lander into the atmosphere of Venus, reducing its mass to 2.5 tonnes, and then use the gravity of the planet to re-aim itself for intercepting the comet.

The cluttered appearance of the Vegas reflected the chief difference from the Giotto mission. The Soviet spacecraft would pass through Halley at a distance of about 10,000 kilometres from the nucleus, where the dust hazard was supposedly much less than at Giotto's nominal 500 kilometres. The Vega designers used wing-like panels of solar cells and they allowed booms, masts, antennas and instruments to protrude indiscriminately. Unlike Giotto, the Vegas did not spin but were fully stabilized, making the operation of the television cameras simpler.

The French had a big hand in the Soviet mission, which was international in its scientific instrumentation. A Marseillaise group was prominent in the Vega camera team, together with Jacques Blamont and Jean-Loup Bertaux of Verrières-le-Buisson. French observatories introduced infra-red telescopes of a kind not carried in Giotto. Hungarian, Austrian and American experimenters made contributions to the Vega payloads, as well as the Soviet scientists led by Roald Sagdeev.

Some Western European experiments that failed to find a niche in Giotto turned up in Vega, including a neutral-gas mass spectrometer from Katlenburg-

Lindau and a plasma-wave experiment from the European Space Agency's own scientists at Noordwijk. The most privileged researcher was Jochen Kissel of Heidelberg. With three nearly identical dust mass spectrometers in Vega-1, Vega-2 and Giotto, more heavily built in the Soviet versions, Kissel had forty-eight kilograms of experimental hardware taken for him to Halley's Comet.

Though full of admiration for their Soviet friends, Western scientists and engineers were bemused by their hands-on approach to building and design. You chose your place amid the Vega jumble by seeing where your instrument might fit. If you were fussy, you might offer up a mock-up of your instrument to the spacecraft. When the flight model was ready, the craftsmen would bolt it on and wire it up. The informal Soviet ways were often quicker and more flexible than the meticulous pre-planning of US and Western European spaceflight. Experiments could be added or subtracted almost up to the last minute. With their powerful launchers the Soviets could afford to be generous with mass, but they were casual about cleanliness too.

Eighteen days after Vega-2 departed into the Solar System, fingerprints and all, a Japanese spacecraft set off for Halley. Its name was Sakigake. Japan was the first country to go firm on a Halley flyby, back in 1979, but it had to develop a new solid-fuel launcher and build a new tracking station for its first deep space mission. With two spacecraft smaller than Giotto, the aims of the mission were sensibly modest. Sakigake ('Forerunner') would pass the comet at 7 million kilometres, chiefly observing the solar wind.

Its follow-up Suisei ('Comet') would be launched after Giotto, in August 1985. Its flyby would be at about 200,000 kilometres, with an ultraviolet telescope as its main instrument. Unlike the Soviets and Europeans, who concentrated on the hours of close approach, the Japanese scientists planned to watch Halley from afar over a period of several months.

A manmade comet appeared in the sky high over the mid-Pacific at Christmas 1984. In a joint US-German-British space experiment, a canister of barium and copper-oxide powder was vaporized in the solar wind. A yellow-blue flash quickly turned purple and within a few minutes the solar wind had entrained the material to create a plasma tail stretching for 15,000 kilometres. It showed waves and kinks similar to those seen in real comets' tails.

Curing the problems of the all-important despin motor, for Giotto's radio dish, took an agonizing year. Its tendency to stick, noted in the electrical tests at Bristol, seemed to be due to a combination of a faulty electronic control system and slight imperfections in manufacture. These were put right. The despin mechanism was finally deemed to be qualified for the mission after tests carried out on the spare model of Giotto, at the friction laboratory in England. It jittered a little in its rotation, which caused further jitters among the engineers, but a careful analysis showed that the performance of the spacecraft would not suffer.

Merriment tinged with hysteria infected those who were working long hours all across Europe to have the spacecraft ready to launch. In an emergent pattern of luck, Giotto was blessed with good fortune about everything that really mattered, while a spirit of mischief haunted the cosmic enterprise and caused alarms.

In March 1985, Intespace caught fire with Giotto in it. The building's lights went off and so did the pumps that raised the air pressure to stop harmful specks of dust creeping into the clean room. Terry Harris of British Aerospace took charge of the rescue. Working by flashlight, he sealed off the area as best he could. As luck would have it, Giotto was waiting to go on a road journey. The rescuers quickly stowed it in its protective container and took it on its trolley to an exit. There they waited, ready to roll Giotto out of the door, until the fire was put out.

The spacecraft was bound for the IABG facility in southern Germany, for overall tests of its magnetism. It travelled on a deep loader and the trip ran true to form, with a diversion at road-works putting Giotto on an autoroute where it was not supposed to be. The police gave chase, and thought that the British Aerospace men accompanying the deep-loader had hijacked it. A snow plough had to extricate the spacecraft from a snowstorm in the Alps. On the way back, after the tests, the truck broke down and the spacecraft came to rest in a farmer's field. French soldiers helped Giotto on its way to Halley's Comet by unceremoniously craning it on to a tank transporter and delivering it to Toulouse.

The European Space Agency formally accepted Giotto from British Aerospace on 22 April 1985. The Bristol company's contribution to the mission was far from over, and its engineers' intimate knowledge of Giotto would be available at every critical stage. A week later, a Boeing 747 cargo plane rose from the airport at Toulouse and headed across the Atlantic towards tropical South America, carrying Giotto and many pallets of supporting equipment. A two-month launch campaign was about to begin, at Kourou in French Guiana.

With its steam-bath climate and the former prison of Devil's Island as the local tourist attraction, Kourou was very different from the neat streets of home. Europe's main space base was sited close to the Equator to maximize the free boost from the Earth's rotation, and Kourou offered an eastward flight path over the ocean where the spent rockets could splash down without killing anyone. So the teams from ESTEC and British Aerospace, and representatives of each of Giotto's scientific experiments, had quickly to adapt to a rain-forest environment and a French-colonial way of life.

David Dale insisted on changing his hotel when he saw a tarantula on the dining-room wall, although colleagues wondered why he imagined that venomous spiders might be choosy about their lodgings. There were a lot of guns about too. The amiable Giotto folk befriended both the local people and the Foreign Legionnaires who guarded the launch base. The French expatriates' notion of the right proportion of rum in a punch was in reality the worst threat to health.

Uwe Keller's camera was, as usual, missing from the spacecraft. The flight model of the camera was delivered to Toulouse earlier in the year in time for the magnetic tests on Giotto at Munich, but at the end of March the camera team asked for their instrument back. They wanted to calibrate its scientific performance at their Belgian installation in Liège. During the tests the camera's rocking mirror seized up.

Nothing would do but to take the camera back to Katlenburg-Lindau and dismantle it. Keller's team worked day and night to try to find out what was wrong, without success. They went with the equipment to the tribology laboratory in England, where leading experts in sticky machinery were equally baffled. After six weeks, when the situation was desperate, the mechanical engineers resorted to a traditional, no-nonsense solution. They machined the axle of the mirror to make it a looser fit. The precious camera travelled to Kourou on an airliner passenger seat, and rejoined Giotto on 17 May. Even then there was a problem with a microswitch, which had to be corrected in the clean room at the launch base.

Another late arrival at Kourou was Hans Balsiger's ion mass spectrometer. Up to the last moment, the team at Bern had been swapping components between the flight model and a spare to achieve the best possible combination. Then, during a late test at Kourou, half of the Bern experiment failed to switch on.

Two engineers representing Balsiger were agitated. They wanted to rip the defective part out of the spacecraft and repair it. Dale disliked doing anything in a rush. He always asked people to think carefully before they 'waded in' at the risk of doing more harm than good. It was a Friday afternoon, and a long-weekend break was just starting for the teams at Kourou, who had been working up to fifteen hours a day.

Dale said No to John Credland and Credland said No to the Swiss. On the phone to Hans Balsiger in Bern, Credland said he would not look at the task until

David Dale

David Link

Hans Keller

Susan McKenna-Lawlor

Henri Rème

Rüdeger Reinhard

the following week. If there were any suggestion from Bern by then, about what might be wrong, he would be glad to know.

'How can we just walk away from that for three days?' asked one of the other experimenters, sympathetic to his colleagues' plight.

'It will be all right, you'll see,' said Dale.

Holiday-time in Kourou included expeditions into the forest, silly games and unlimited quantities of the most diluted rum punch ever made in French Guiana. The only technical question was which was funnier, the foundering of John Credland's water scooter or the confrontation in the jungle between a deadly snake and a lady physics professor from Ireland.

At Bern University, 7000 kilometres away, Balsiger and his team spent a miserable weekend with the simulator of their experiment and spare hardware. As Dale expected, they managed to figure out where the fault in their instrument probably lay. To put it right was a simple resoldering task.

Balsiger suspected that Credland might hold up the work until it was too late to fit in before the launch. Then the instrument would go off to Halley and could fail to work on arrival. So Balsiger planned to take the issue up with Roger Bonnet in Paris and, if need be, with the Swiss delegation to the European Space Agency. On Tuesday, as soon as everyone was back at work at Kourou, Balsiger was on the phone to Dale.

'If I don't get a positive answer in two hours . . . ,' Balsiger began.

'Of course they can fix your experiment,' Dale said.

Balsiger would always believe that Credland was the bad guy, and Dale the reasonable manager who let his people attend to the instrument at last. In fact, as always, Credland and Dale together knew exactly what they were doing. Theirs was a tough regime, though, and Balsiger's integration engineers both fell ill on their return from Kourou to Europe.

The Ariane launcher went on to the pad for careful checkouts. As David Dale wanted to minimize any trouble with the rocket during the countdown, he insisted on a full rehearsal of the troublesome task of filling the third stage of the rocket with the liquid oxygen and hydrogen used as the propellants. The only time available was at the weekend, and the launch team stipulated that Dale too should attend the practice. As the propellants were potentially explosive, the pad area was sealed off and all concerned went to the launch-control bunker.

The automatic sequence for filling failed, so the Frenchman in charge gave a sequence of commands to his crew to do the fill under manual control. Dale was satisfied that, even if the computer were not fixed in time, the launch could still go ahead. Then the oxygen and hydrogen were vented. The whole operation took nearly twenty hours instead of the seven or eight originally envisaged. The launch crew sat back, yawning at their consoles.

'What happens now?' Dale asked the Arianespace mission manager responsible for seeing Giotto safely launched. 'Why aren't we all going home?'

'We have to wait six hours more. It's still dangerous out there until the gas has dispersed.'

To Dale's surprise, the massive fireproof doors of the bunker then opened and revealed a fire engine backed up to it. The *pompiers* offloaded chicken and wine for a party in the bunker.

'I thought there were restrictions on the pad,' Dale said.

'Ah, not for the *pompiers*.'

Word came from Moscow that the Vega spacecraft had dropped their balloons and landers into the atmosphere of Venus and successfully altered course for Halley. With two identical spacecraft, the Soviets could afford to lose one of them. There was only one flight-qualified Giotto and the war with possible gremlins of technical malfunction in the spacecraft or the launcher was correspondingly fierce. The unspoken thought in everyone's mind was that all their work could end up as scrap metal on the bed of the Atlantic Ocean.

Filling the fuel tanks for Giotto's thrusters, which would control the spacecraft during its cruise to Halley, was one of the last tasks at Kourou. The hydrazine fuel was highly poisonous as well as potentially explosive, and technicians dressed as if for chemical warfare put it into four bulbous tanks. These supplied four pairs of thrusters evenly spaced around the craft. Each thruster was a miniature rocket motor, in which a catalyst would decompose the hydrazine to produce a jet pushing in a selected direction.

When Giotto was in place on top of the rocket, shortly before the launch, mischief reappeared from a quite unpredictable source. Japanese visitors taken up the launch gantry started photographing the spacecraft using flash bulbs. This was an unthinkable assault by photons on the sensitive solar cells and there was uproar when the Europeans heard about it. Dale threatened to cancel the mission and send the bill to the company hosting the visitors.

To keep its appointment with Halley's Comet, Giotto was to depart from Kourou no sooner than 11.13 zulu on 2 July. The 'zulu' time of space operations and aviation was more familiar to scientists as Universal Time and to laymen as

Greenwich Mean Time. For reasons going back to the choice of the zero of longitude, the clocks for operations encircling the Earth or reaching out into the depths of space were set to the solar day in an eastern suburb of London.

The launch window lasted twenty-two days. The countdown began on 30 June, either to perform the launch on the first possible day, or to allow the maximum time to sort out any problems in the event of a delay.

The launch controller needed inputs from the various teams, confirming that the rocket, the spacecraft, the launch range of Kourou, and mission control in Europe were all ready. David Dale sat at one end of the row of consoles in the bunker. He had a permanent phone link to the European Space Operations Centre at Darmstadt in Germany, which would take charge of Giotto after the launch. ESOC in its turn was monitoring the condition of its mission control centre and the ground stations around the world that would track Giotto after launch.

Another line connected Dale with his project team that was watching, in a separate room, the telemetry coming down from the spacecraft atop the rocket. Late in the countdown, the project team announced a crisis.

'The temperature of the spacecraft is dropping like a stone. They seem to have forgotten to put in the insulation.'

Dale pictured the scene inside the fairing in the nose of Ariane. The spacecraft sat on an adaptor that mated it with the rocket, just above the hydrogen fuel of the third stage. If the protecting layer was missing, Giotto was looking directly at a tank filled with extremely cold liquid hydrogen. No wonder it felt chilly.

Dale did not want to damage Giotto. Most at risk was the mechanism that would cover the aperture of the onboard Mage motor after it fired. But to halt the launch and then take off the spacecraft for the insulation to go in would require, he calculated, at least two weeks of his three-week window.

'How far can we go down?' he asked the spacecraft engineers.

'Minus 25.'

'Where are we now?'

'Minus 20.'

'There's half an hour to go. Let me know when you get to minus 30.'

Dale was glad not to hear from the project team again, on this subject, even though there was a brief hold in the countdown at minus six minutes. Eventually the gantry arms fuelling the third stage swung away. The first stage ignited and with a great roar Ariane lifted Giotto into the tropical air just ten minutes after the earliest permissible moment of launch.

Everyone was still tense. The rocket engineers had experienced failures be-

Giotto, on top of the Ariane rocket, shortly before its launch at Kourou, French Guiana.

fore. The Giotto teams scarcely breathed for quarter of an hour as the stages of the Ariane fired in turn. The dip manoeuvre to shed energy occurred. The spacecraft finally separated. Giotto's luck had held. Unknown to most of the watchers, the telemetry showed an incipient fault in the rocket's third stage. Two months later, because of the same fault, the next Ariane launch failed spectacularly in front of the president of France.

When Giotto parted from the third stage, fifteen minutes after lift-off, it was passing north of the mid-Atlantic island of Ascension. It then had to go into a series of pre-programmed operations to switch itself on, and continue eastwards around the world before coming in contact with the first ground station, in Kenya. Everyone still waited in dread until Dale's Darmstadt phone flashed.

'We have telemetry, Dave.' Giotto was transmitting.

The celebrations began at Kourou, with the *pompiers* catering as usual. The scientists could still look forward to an eight-month mission, but for most of the ESTEC team Giotto's launch ended the busiest time of their lives. The human cost of sending a spacecraft to Halley's Comet included further physical illness and nervous breakdowns, when the reaction set in.

The day of Giotto's launch would be seen, with hindsight, as one of history's bright dawns. It was on 2 July 1985 that a new Soviet leader, Mikhail Gorbachev, exchanged his hard-line foreign minister Andrei Gromyko for the liberal Eduard Shevardnadzhe. This was the first clear signal of changes that would end the Cold War, dispel the spectre of nuclear conflict between the superpowers, and free the peoples of the Soviet bloc.

Between the Rhine and the woods of the Odenwald, just south of Frankfurt, the small and friendly city of Darmstadt was one of the places where the German chemical industry took shape. It became Europe's chief channel to outer space when chosen as the site for the European Space Operations Centre (ESOC) and its complex of control rooms, computers and offices for operating the European Space Agency's satellites.

Giotto's flight operations director was a burly Welshman, David Wilkins. He learned his trade with NASA and the Gemini series of manned spacecraft in the 1960s. Like the conductor of an orchestra, Wilkins was surrounded by a semicircle of specialists sitting at their consoles. Some were responsible for flight dynamics, the ground stations, or all the hardware and software required to support the mission. Others were operations engineers who commanded the spacecraft and kept an eye on its health. Andrew Parkes, who shared his name with the

Australian radio telescope, shouldered the main day-by-day responsibility for flying Giotto to Halley's Comet, as spacecraft operations manager.

It was a small team by American standards. NASA would have up to ten times as many people in mission control as the European Space Agency thought strictly necessary for a given task. When all went well, low manning levels paid off handsomely in cheaper missions. But trouble with the spacecraft or a change of plan could put an enormous pressure on a small team, which could find itself working double shifts for days on end.

'One of these days someone will make a very expensive mistake,' Wilkins would say, suggesting that European parsimony sometimes went too far.

Esoc had been involved with Giotto from its conception. In the winter of 1979-80 the basis for discussing a European comet flyby came from a future-missions team at Darmstadt led by a dapper Swiss orbital mechanics specialist, Walter Flury. It started from predictions of the motions of Halley and the Earth, in their orbits around the Sun, during the comet's visit of 1986. The calculations specified the launch date, the interplanetary route, the angle at which Giotto would converge with the comet, and the date and time of day of the encounter. The mass of the spacecraft in relation to the launching energy, the relative speed of the comet's dust particles, the necessary tilt of Giotto's radio dish, the choice of Parkes in Australia as the main downlink – all of these features of the mission arose from the early analysis at Darmstadt.

A flight dynamics team took over, well in advance of the launch, to study the details of the orbit to Halley, the communications opportunities from different parts of the Earth surface, and Giotto's correct attitude, or orientation in space, at each stage of its cruise to its appointment with the comet. While the flight dynamicists provided the mathematics, the spacecraft operations team were engineers by training and disposition.

You could learn to fly a spacecraft with practice, like driving a car, but a spacecraft controller had to know far more about what went on under the bonnet. When Howard Nye, a young Englishman, was appointed to the Giotto flight operations team more than two years before the launch, he found himself back in Bristol, where he had worked as a spacecraft engineer. He reviewed with British Aerospace the attitude and orbit control systems, and the questions of power and temperature requiring constant attention during the mission.

'You have to know exactly what the spacecraft is intended to accomplish,' Nye would say to anyone who pushed the car-driving analogy too far. 'And you have to use the facilities or even design new facilities to allow this to happen.'

Long before the launch, the flight control team began rehearsing the operations using a computerized simulator at Darmstadt. A phantom Giotto received commands, altered its behaviour in its phantom orbit, and returned appropriate telemetry with a high degree of realism. As with the flight simulators used by airlines, the controllers could also practice reactions to emergencies, when red lights indicated departures from the norm.

Rehearsal gave way to reality. The flight control team shared the anxieties of the launch, but with the adrenalin rising. The separation from the launching rocket marked the end of Kourou's responsibility for Giotto's welfare. For better or worse, the spacecraft was in Darmstadt's hands.

For a day and a half after launch, Giotto went around the Earth, as planned, on a highly eccentric orbit. To enter interplanetary space, Giotto had to use its onboard Mage motor when it was at its closest to the Earth. The main uplink station for sending commands to the spacecraft was at Carnarvon in Western Australia, but Giotto would be travelling so fast that the dish could not swing quickly enough to follow it. The controllers therefore pre-programmed the firing of the Mage.

Exactly on cue, the motor fired and consumed its solid fuel in less than a minute, leaving Giotto 374 kilograms lighter as it broke away from the Earth to rush off into the Solar System. The first confirmation came from changes in the radio signals from the spacecraft, which were shifted in frequency by its increased speed.

Giotto's route to Halley was an orbit roughly similar to the Earth's, but dipping closer to the Sun. Travelling faster, the spacecraft would cut off a corner, so to speak, to intersect the orbit of the comet close to the Earth's orbit but far ahead of the home planet. By a paradox of celestial mechanics, Giotto lagged a little behind the Earth after launch and remained fairly close for the first few months.

After tracking it for some days, the controllers could see that their calculations and the Mage burn were both so accurate that no major course correction would be needed to ensure a close encounter with Halley eight months later. Any small adjustments could be left until the astronomers had more accurate news on the comet's movements. Much of the hydrazine thruster fuel already seemed superfluous.

Before anyone could breathe easily, one more step was essential. The early operations close to the Earth used the small look-anywhere antenna for communications to and from the spacecraft. Giotto's radio dish, essential for distant communications, had to be pointed at the Earth and switched on. Would its troublesome despin motor perform as required? It did. The motor worked flawlessly throughout the mission. A slight waving of the radio beam occurred as a result of electrical interference in the star mapper serving the Earth-pointing control system. As there were other ways of controlling the aim of the radio dish, this never became a serious problem.

Flight dynamicists remained on hand throughout the cruise to Halley, to advise on the rare adjustments to the orbit and the frequent corrections to the spacecraft's attitude. Giotto's thrusters executed the manoeuvres. When commanded to burn, two thrusters could fire forwards, two backwards and two sideways, and a fourth pair, directed tangentially, controlled the rate of spin. Alterations to the spin, attitude and speed were interconnected in ways that sometimes contradicted intuition, like the behaviour of a gyroscope. The electronic flight simulator remained in use, to predict how the spacecraft should respond to a given burn or combination of burns by the thrusters. When the commands went out to the real Giotto it did just what was expected.

That a spacecraft should always perform as the designers intended was not taken for granted. 'Some are like Friday cars, with something always going wrong,' Wilkins remarked to David Dale. 'Giotto is a good spacecraft.' Dale was pleased but not surprised. He knew that the people who built Giotto left nothing to chance. The user's manual that British Aerospace supplied with their spacecraft also earned special praise.

For all that, Giotto was no spacecraft to fire and forget. The orientation of its spin axis in relation to the Sun was critical, to maintain power from the solar cells and avoid overheating or overcooling of any parts of the spacecraft. But the radio dish, with its permanent squint, had to point at the Earth.

The operations manager had a narrow band of permissible attitudes for the spacecraft that changed from week to week. The spacecraft always tended to drift out of alignment as it moved forward in its orbit. Besides adjusting the attitude directly, the controllers had to keep coaching Giotto's systems of autonomous control with updated instructions on safe attitudes to adopt in an emergency.

The flight control team went into a skeletal mode of operation for the long cruise out to the comet. As for the project management, Dale and his team began studying a new scheme for the group of scientific satellites called Soho and Cluster, for research on the Sun and the solar wind. This study kept the team in being. Dale and the others could always break off to attend to Giotto business, or take part in formal tests and rehearsals.

For much of the time Rüdeger Reinhard was the project management's chief representative at Darmstadt. There, the Giotto scientists were covering ever-larger areas of ESOC's floor space with equipment that they wanted for the encounter. Soon they would mercilessly exploit the good nature of the mission controllers, to do more and more science en route.

The Japanese dispatched their Suisei spacecraft towards Halley's Comet, on 18 August, seven weeks behind Giotto. By steering closer to the Sun it would overtake Giotto and arrive in the comet's vicinity five days earlier, and slightly ahead of its own companion, Sakigake, launched back in January on a wider orbit. With two Japanese, two Soviet and one European spacecraft all bound for their several encounters with Halley in March 1986 the international fleet was complete.

On 11 September 1985 leaders of all these missions met at NASA's Goddard Space Flight Center near Washington to see the Americans upstage them all, by flying a spacecraft through a comet for the first time ever. ICE, or International Cometary Explorer, formerly International Sun-Earth Explorer number 3, had found its target: Comet Giacobini-Zinner.

'Getting there was half the fun,' said Robert Farquhar of Goddard, who had contrived the mind-boggling series of five swingbys around the Moon that sent the elderly spacecraft on its way. At the time of the last of these manoeuvres the target was not in sight. Fortunately Giacobini-Zinner showed up on its 6.5-year orbit exactly as required for its interception. Although the mission was led and stage-managed by the Americans, International Cometary Explorer deserved its name. Of the seven main experiments operational during the comet encounter, three had principal investigators from Europe – Munich, Paris and London.

Giacobini-Zinner was a far less active comet than Halley. ICE went through its tail, 7800 kilometres behind the apparent position of a nucleus. As ICE had no dust shield the Americans feared that the comet might smash it, but it survived essentially unscathed.

Although not designed for a comet mission, and having no camera, ICE was able to measure the particles and magnetism of the solar wind. It had a limited capacity for chemical analysis. Dust impacts had to be inferred from disturbances to instruments designed for quite different purposes. Yet this improvised venture led to discoveries, as ICE observed at first hand the battle between the atmosphere of the comet and the solar wind.

'Maybe we have to switch on our experiments earlier than planned,' Rüdeger Reinhard said. As Giotto's project scientist watched the data arriving at Goddard he was startled that turbulence caused by the comet showed up farther out than predicted. On the other hand, Giacobini-Zinner had no clear bow shock, where the solar wind was expected to slow down abruptly at the edge of the comet's domain.

The encounter confirmed a theory in comet science. The magnetism of the solar wind switched in direction as ICE crossed from the near side to the far side of the comet's tail, just as a Swedish astrophysicist, Hannes Alfvén, had predicted eighteen years earlier. A chemical instrument indicated the presence of water in the tail, as expected if the nucleus were an icy body.

'It's another one of those firsts that this country enjoys making,' the NASA administrator remarked of the Giacobini-Zinner encounter. The US Post Office celebrated the event with special postmarks. ICE continued on its interplanetary travels to make a very distant flyby of Halley.

Another American spacecraft, Pioneer Venus, was in orbit around the planet Venus. When it turned its ultraviolet telescope towards Comet Giacobini-Zinner, Pioneer Venus showed that the hydrogen cloud around the comet was at least ten million kilometres wide. Pioneer Venus would look at Halley too. The Americans had other programmes and collaborations in hand for observing Halley from spacecraft operating near the Earth, including International Ultraviolet Explorer and Solar Maximum Mission. Space scientists were preparing special Halley-observing instruments to fly in space shuttle missions early in 1986.

Professional and amateur astronomers all around the world watched Halley's Comet from the ground. Since first seen far off, three years earlier, the comet had drawn much closer to the Earth and Sun. The warming sunlight was provoking ever-increasing emissions of dust and gas, and although the Earth's position in its orbit left the comet much farther away than usual, any amateur astronomer worthy of the name could find it. The more skilful ones contributed thousands of measurements, photographs and drawings to the records of the International Halley Watch.

Astronomers, book publishers, newspaper editors, impresarios, manufacturers of memorabilia, and occasional cranks all joined in whipping up public excitement about Halley. They had to cope with the awkward fact that it would be an unusually poor spectacle to the unaided eye. 'It will not be a lazy man's comet,' admitted the president of General Comet Industries Inc. of New York, which sold shares in Halley's Comet.

Comet pills on sale in the US, meant as a joke, recalling the quacks of the 1910 apparition. An organization called the Grace of Jesus Christ Crusade did not seem to be jesting when it put out a broadsheet which declared that Halley's Comet was the herald of the world's end. Its authors updated the traditional list of comet-induced natural disasters, famines and epidemics by adding herpes and AIDS.

More soberly, in the discoverer's own country, the Halley's Comet Society arranged for a plaque commemorating Halley and the Giotto mission to be put up in London's Westminster Abbey. A National Astronomy Week coincided with the best time for viewing the comet from Europe, in mid-November 1985, although you still needed a telescope or very good binoculars to see it.

Smart operators exploited the comet's poor visibility from northern lands to arrange Halley-viewing expeditions to the southern hemisphere in 1986. A Japanese comet camp would spring up in Australia, and the locals in South America, South Africa and various oceanic islands would find themselves hosts to neck-craning tourists.

CHAPTER SIX

THE FIRST PICTURES

UWE KELLER noted that Giotto's wobble was very slight, at only one-hundredth of the specified maximum. The five kilograms of hardware and countless man-hours of programming that went into his camera's computer turned out to be way beyond what the job required. David Dale declined to discuss the might-have-beens. If the spacecraft performed better than its specification, that was a matter for congratulation not complaint.

On 13 September 1985, when Giotto had been in space for ten weeks, the camera turned towards the star Vega for a test of its optical performance. Keller was nervous. The camera was never tried out properly on the ground, as a complete system producing real images, because of the last-minute problem with the sticky mirror. An unnoticed error in design or construction could make it useless.

Sharp images of the star, transmitted from Giotto, came as a relief. Two weeks later, tests with the planet Jupiter were successful too. But a light-sensor, used for triggering the camera in one mode of operation, failed to work. Perhaps Vega and Jupiter were too faint. The way to check it was to look at a brighter object.

The Earth was the only candidate. Keller wanted to turn the camera on it anyway, while Giotto was still relatively close. It made an ideal test object. The Earth seen from 20 million kilometres resembled in size and brightness the heart of Halley's Comet seen from 20,000 kilometres, as it would appear from Giotto five minutes before the closest approach.

But Dale hated the idea. The attitude experts at Darmstadt noted that if Keller turned the camera towards the Earth, its protruding tube would cast a shadow on Giotto's solar cells. Suppose the camera became stuck in that position, for any of half a dozen reasons. The spacecraft would suffer a permanent loss of part of its power supply and its performance at the encounter might be blunted.

Rüdeger Reinhard had tried for years to keep the peace between his friends the project manager and the camera principal investigator, but he saw a new fight brewing. Dale pointed out to the camera team, not without mockery, that in one of their proposed operations the camera would have swept around until it was looking straight into the Sun – thereby destroying its delicate light-detectors. The manoeuvre could be done another way, but the argument about whether the Earth should be observed at all had to go for adjudication to Roger Bonnet, the European Space Agency's director of science.

'If you want me to authorize this,' Dale said, 'you must order me to do it.'

On Bonnet's recommendation, the director general gave the order. After further tests of all of the camera's hardware and software, Keller was at last allowed to take peeps at the Earth, on 18 and 23 October 1985. Its diameter spanned only twenty-seven detector elements in Giotto's camera, but large cloud patches over Australia, central Asia and Antarctica were discernible. These were also visible in images from a Japanese weather satellite. The camera gave no trouble, but ESOC's own computer system went down during the operation and the controllers spent a frenetic quarter of an hour dictating their commands by phone to the ground station at Carnarvon in Australia.

Although the camera team's relations with the project management remained strained, it enjoyed fine cooperation throughout the cruise from mission control. So did the other experiments. According to the original mission profile, they were to be switched on only for a few tests and rehearsals, before their arrival near the comet. The first exceptions were Fritz Neubauer's magnetometers and Susan McKenna-Lawlor's energetic-particle telescopes. Their low data rates made it practicable to record their measurements on board, and relay them whenever mission control was in contact with Giotto for other reasons.

Then the proposed camera operations multiplied. Other experimenters, with instruments capable of detecting the solar wind through which Giotto was cruising, asked for observing time too. There was a functional spacecraft out there, so why not use it? Dale relented and arranged with the downlink station at Parkes and with the NASA Deep Space Network for the spacecraft to make scientific observations two or three times a week. There was no increase in the manning levels at mission control. Howard Nye, as a member of the operations team with special responsibility for the experiments, found his workload increasing from a nominal couple of hours to fourteen hours a day, and not just now and again but routinely throughout the cruise.

The daily round at Darmstadt was set by the hours at which the uplink and downlink stations in Australia were facing the distant Giotto. In the unceasing sunshine of interplanetary space, the spacecraft knew nothing of night and day. By the end of December it had drawn well ahead of the Earth in its solar orbit. The range was 75 million kilometres and increasing at more than a million kilometres a day. The geometry of the Solar System then made the time for Giotto-talk the early morning in Australia and the evening at Darmstadt.

When the scientists told Nye what they wanted Giotto to do, he had to write and carefully verify the coded commands. The camera was especially quirky to deal with. Whenever it was switched on, it would spontaneously start searching for a comet. If the camera team wanted to look at a planet or a star, it had to draft complex codes to override the camera's comet- hunting instinct. The camera had its own electronic simulator for testing the commands for possible mistakes. Nye then had to check them line by line, and consider possible effects on the health of the camera and the spacecraft as a whole. Often that meant sitting up all night.

As project scientist, Reinhard was deeply sceptical about this cruise science. All that Giotto could observe was either the normal solar wind, for which better instrumented spacecraft existed, or distant visible objects that the camera could see less well than telescopes on the Earth. In Reinhard's opinion, any results would be too meagre to justify the effort. He kept his thoughts to himself. If Nye and the others at Darmstadt were crazy enough to gratify his scientists' whims, he did not wish to dampen anyone's enthusiasm. And it was all a rehearsal of a kind, for Halley's Comet.

In January 1986, a matter of weeks before Giotto was due at its target, the US spacecraft Voyager-2 made its encounter with the planet Uranus. NASA's Deep Space Network was stretched close to its limits, to harvest images of Uranus and its moons transmitted from a distance of nearly 3 billion kilometres. Thanks to the close relationship that had developed between the Giotto mission and the headquarters of the DSN at the Jet Propulsion Laboratory in California, the Australian radio telescope at Parkes, adapted for Giotto, provided an extra ground station for Voyager.

David Dale, Giotto's project manager, and David Wilkins, the flight operations director, were both visiting JPL at the time of the Uranus encounter. Dale was woken at 3 a.m. in his Pasadena hotel by a phone call from Darmstadt.

'We've lost Giotto.'

'Don't be silly. You know where it is,' Dale said, and went back to sleep. Ten minutes later the phone buzzed again.

'We've lost Giotto's signal.' Mission control had run out of time while sending a set of commands to Giotto. From day to day the spacecraft received 'safety coaching' on what to do in the event of an emergency. But on this occasion the transmission had been delayed. Before the commands were complete, the spacecraft had already drifted into an orientation disallowed by previous instructions.

Giotto automatically swung into its safest attitude, as if to say, '*Che ridicolo!* Until you arrange yourselves I'll attend to my own affairs.' It made sure that its solar cells were nicely caressed by the Sun, and its temperature controls optimized. The radio dish no longer faced the Earth and communication was lost. In a day or two, Giotto ought to go into another automatic routine to look for the Earth and then orientate its dish towards it in the hope of hearing something. That operation had never been tested in space, and in any case inconsistencies created by the uncompleted coaching would puzzle the spacecraft.

Dale sensed danger, so he phoned Wilkins, who was staying in another hotel. The two men met in the small hours to see what they could do. The only sure procedure would be to give Giotto direct commands to correct its attitude. The European uplink station at Carnarvon in Australia did not have a strong enough beam to send orders via the spacecraft's auxiliary, look-anywhere antenna.

'We'll have to ask DSN for help,' Wilkins said.

Seeking assistance from the Deep Space Network while it was engrossed with Voyager-2 was like interrupting the surgeon during an open-heart operation, but Dale and Wilkins clenched their teeth and made their request. The Americans were very obliging. It was a test of their resourcefulness.

The snag was that Giotto and Voyager-2 lay far out in space on the same side of the Earth. The big DSN dish at Goldstone in California, which could point at Giotto, was handling Voyager-2 at the start of its encounter with Uranus. But the American spacecraft was westward of Giotto and would come into view from Australia while Giotto was still accessible from Goldstone.

The NASA engineers therefore transferred the link with Voyager-2 to the DSN station near Canberra as soon as possible, and left Goldstone free to turn to the European spacecraft. In loud commands, relayed from Darmstadt, Giotto recognized the voice of authority and slewed back into normal communication.

Voyager-2's pictures of Uranus were of special interest to comet scientists, who suspected that comets were born in the same part of the Solar System. The images of the moons of Uranus gave Uwe Keller cause for thought. They included small ones as dark as soot. If Halley's nucleus were as sombre as those, would Giotto's camera ever find it, amid the fireworks of the comet's atmosphere?

The explosion over the Florida coast on the chilly morning of 28 January 1986, which destroyed the launching rocket of the space shuttle Challenger and killed the seven astronauts aboard, sent a wave of horror and sorrow through the world's space agencies. Several individuals in the Giotto team had worked with the US manned spaceflight programme and knew what their friends were suffering. The European Space Agency was involved with the shuttle, through its own Spacelab that fitted into the cargo bay, and through several joint US-European missions that relied on shuttle launches.

Immediately affected were US projects for Halley's Comet. The astronauts in Challenger itself were supposed to deploy a free-flying experiment, Spartan Halley, for observing the comet by ultraviolet and visible light, and to operate a special Halley camera called Champ. Another package of ultraviolet instruments, Astro-1, was meant to go in the next shuttle flight. The fact that would-be comet observers had perished in their mission cast a shadow over the International Halley Watch.

It was a bad day for space science in general. NASA had advertised its reusable spaceship as the cheap way to carry payloads into space, but rising costs of the shuttle had already curbed the US space science programme and helped to stifle the International Comet Mission. Even before the Challenger disaster, several major scientific spacecraft adapted to shuttle launches were affected by delays in the programme. They included the Hubble Space Telescope, the European-built solar probe Ulysses and the European 'retrievable carrier' for materials science called Eureca. The grounding of the remaining shuttles, while the fault in the booster motors was identified and cured, postponed them *sine die*.

On 8 February Giotto registered a shock from the Sun. It was the most striking event in all the months of cruise science and was the more surprising because this was in a quiet phase of the eleven-year cycle defined by the count of dark sunspots. Early in February the face of the Sun suddenly became spotty around two active regions, and observers on the Earth saw several major solar flares, electromagnetic explosions in the Sun's lower atmosphere. An exceptionally big flare occurred on 6 February.

A burst of energetic particles raced outwards from the Sun, overtaking the slower particles of the normal wind at a shock wave. When it slammed into the Earth's magnetic shield on 8 February, the particles caused the most violent magnetic storm ever recorded. Observatories in the US saw the local direction of the magnetic field vary temporarily by up to eighteen degrees. People in England, the Netherlands and Germany witnessed unusual auroral displays in the northern sky. Radio communications suffered as electrons in the ionosphere, the natural radio reflector in the upper atmosphere, first doubled in numbers and then almost vanished.

Plasma analysers in a satellite near the Earth registered the arrival of the energetic particles, and far away across the Solar System the Johnstone plasma analyser aboard Giotto was operating when the shock arrived at 02.36 zulu time on 8 February. Swift nuclei of hydrogen and helium pounded its detectors. In just a few seconds, the speed of the particles jumped from 370 to 430 kilometres per second. During the next two days the solar wind showed further erratic behaviour, including a surge in the helium content. At one point its speed of the particles touched 900 kilometres per second. The Irish EPONA instrument also recorded high particle counts and shocks in every channel.

Scientists at Boulder, Colorado, later simulated the event in a computer. The shock wave spread across the Solar System in a great arc that reached the Earth and Giotto at about the same time. The Soviet Vegas bound for Halley were farther from the Sun, and recorded the shock a little later.

Susan McKenna-Lawlor was an expert on solar flares, and she was worried that a repeat of the event might occur in the same region of the Sun just before Giotto's encounter with the comet. If so, the solar particles might mask the more subtle effects of particle acceleration close the comet itself, which she was hoping to detect. Fortunately the Sun reverted to a quiet state more appropriate to its sunspot season.

Giotto's camera produced its first snapshot of Halley on 4 March. The distance was still 59 million kilometres and the resolution no better than from a telescope on the Earth. But the range was shrinking at 6 million kilometres a day and the camera team obtained further pictures at intervals during the approach. They were not very striking, because there was a lot of noise in the signals.

Proudly leading the Halley fleet, Vega-1 flew through the head of Halley's Comet in mid-morning Moscow time on Thursday 6 March. It passed 8900 kilometres from the bright spot in the centre, supposedly the nucleus, and on Sunday 9 March Vega-2 followed it through, at 8000 kilometres. With Roald Sagdeev presiding, the stage management was excellent. Data relayed from mission control, at the deep space tracking station at Evpatoria in the Crimea, appeared on the consoles of the multinational teams of scientists assembled at the Space Research

Institute in Moscow. Visitors from the European Space Agency formed a vivid impression of how intense the activity would be at Giotto's encounter.

As the first spacecraft to arrive at the comet, Vega-1 had a good chance of finding something new. It detected a bow shock about a million kilometres out, where Halley's atmosphere began to tame the solar wind, but this was expected. More surprising were the many small grains, rich in carbon compounds, turning up in the comet's dust. This was the first major discovery of the international Halley fleet.

The Vega-1 instrument of Jochen Kissel of Heidelberg, very similar to those he had in Vega-2 and Giotto, gave the most detailed results on the composition of the dust grains. Luckily it worked well, because the one in Vega-2 suffered a drop in the power supply when the Soviets corrected the pointing of the camera, and it went through Halley in a less than fully sensitive mode. Later, Kissel and a German colleague would offer a new concept of how life began on the Earth, based on the chemical results from Vega-1 and Giotto.

Apart from those who had experiments in the Vegas, no onlooker in Moscow watched more attentively than Uwe Keller. As he saw the images starting to come in from the cameras, first in one spacecraft and then in the other, the same thought gnawed him. The Soviets might come up with such clear pictures of the Halley nucleus that anything from Giotto's camera would seem unsurprising.

Many exclamations of delight from Soviet spectators greeted the Vega-1 images, but Keller with his expert eye could see that they were poor. Those from the closest approach showed only a bright blob. If there was any detail present it was hard to spot. One instant theory was that the nucleus was hidden in a cocoon of dust. Even Carl Sagan, the American astronomer and a famous popularizer, became reticent when asked to comment on the images in front of a Moscow television camera.

Keller reserved judgment until the Vega-2 images came in. Well, they were different. They showed two blobs. The alteration could be due to a rotation of the Halley nucleus during the three days between the flybys. Certainly there was less dust about during Vega-2's encounter.

A BBC interviewer asked Sagdeev, 'Did you actually see the nucleus?'

'I hope so,' was his reply. 'It doesn't look like solid rock,' Sagdeev said. 'It looks like a sophisticated, maybe double structure. And the edges of this object are made completely obscure because of, probably, a lot of dust streams and jets.'

Keller knew that careful processing of the fuzzy Vega images would reveal more detail, but they gave no vivid impression of what a comet's nucleus was like.

In truth, Vega-1's telescope was out of focus, and Vega-2's images were overexposed. Keller had no feelings of *Schadenfreude* as he boarded the plane in Moscow to return to Darmstadt for the Giotto encounter. He shared a sense of disappointment felt by members of the Vega camera team. And while he allowed himself to hope that his own pictures might be more revealing, he was apprehensive. If the Soviet images looked vague because the nucleus was shrouded in a cloud of nearby dust, Giotto's camera might not see very much, either.

Keller travelled in company with Fred Whipple, the American author of the dirty snowball theory. He too was with the flying circus now on its way from Vega to Giotto. Although nearly eighty, Whipple was as happy as a schoolboy amid the mighty efforts going on to find a comet's nucleus, which might confirm his predictions made thirty-six years before. His mind was still working like youngster's and he spent the flight calculating whether the double blob in the Vega-2 images might be taken at face value. Could Halley have two nuclei loosely bound together by gravity? The numbers did not look favourable.

The first of the Japanese spacecraft arrived in the vicinity of Halley's Comet some fifty-three hours after Vega-1. It was Suisei, which passed 151,000 kilometres from the nucleus on 8 March. For four months, Suisei's ultraviolet telescope had been observing the large cloud of hydrogen surrounding the comet. It saw peaks in the comet's expiration of gas at regular intervals of fifty-three hours. The Japanese scientists attributed the 'heavy breathing' to a rotation of the nucleus, which brought the most active areas to the sunward side.

Around the time of its closest approach, Suisei's instruments observed the turbulence in the solar wind and saw fragments of decomposing water molecules caught up in it. Hits from two relatively large dust grains tilted Suisei slightly and altered its rate of spin, without doing serious damage or interfering with the scientific data-gathering. A Japanese scientist said, 'We don't know whether to call it good luck or bad.'

To suffer palpable hits from comet dust so far from the nucleus, at 300 times the distance intended for Giotto's flyby, was so unexpected it amounted to a discovery. And it heightened anxieties among the Europeans about their spacecraft's safety. Suisei's spin axis changed its orientation by an angle of 0.72 degree. A one-degree change in Giotto's attitude could put it out of communication with the Earth.

Suisei would go on making observations for more than a month. Sakigake, the second Japanese spacecraft, passed 7 million kilometres from the Halley nucleus forty hours after Suisei's passage. From its distant vantage point, this spacecraft observed the solar wind blowing towards the comet and magnetic waves coming from it, apparently stirred by the comet's encounter with the solar wind.

On the evening of Sakigake's flyby day, Monday 10 March, Giotto went through a full dress rehearsal for its encounter with Halley three days later. Staged at the European Space Operations Centre, it involved the mission controllers, the scientists, the ground stations, and everyone else who had a part to play.

It all went just as it would on the night, except that the spirit of mischief that attended the Giotto mission reappeared when Darmstadt suddenly lost communication with its uplink station at Carnarvon in Western Australia. A new link was established in a matter of minutes, and the rehearsal went on. Enquiries showed that Australian builders had put a digger through telephone cables carrying the lines to Carnarvon.

After Vega-2 passed through the head of the comet on 9 March, the international Pathfinder teams rushed to complete their collation of data from the Soviet spacecraft and NASA's Deep Space Network. They had to define as accurately as possible the trajectory of the bright blob or blobs, presumed to be the nucleus, with a view to refining Giotto's aim. Two small mid-course corrections during Giotto's cruise, in August and February, had followed refined predictions made possible by astronomical observations from the ground in the International Halley Watch. But there was still an uncertainty of several hundred kilometres in the position of the nucleus at the time of the spacecraft's arrival.

The Vega team had calibrated the pointing of their cameras by sighting them on the planets Jupiter and Saturn before the Halley encounters. They could therefore fix the bearings of the nucleus in relation to the spacecraft. Changes in direction during the flybys gave the distance off. At the same time, NASA's big dishes worked in pairs and used atomic clocks to measure differences in the arrival times of signals from the Vegas, at widely spaced parts of the Earth's surface. The DSN fixed the positions of the Vegas to an accuracy of about twenty kilometres – equivalent to distinguishing apples on a tree from a distance of 500 kilometres.

Combining the Soviet and American observations, the Pathfinder teams reduced the position of the brightest spot in Halley, when Giotto arrived, to a small ellipse of uncertainty. The spacecraft's miss distance would be controllable to within plus or minus forty or fifty kilometres. Darmstadt had hoped for guidance from the Soviet spacecraft on jets of dust to avoid, coming from the presumed nucleus. Unfortunately, the comet's emissions as seen by the Vegas were too variable and poorly defined for that.

On the afternoon of Tuesday 11 March, two days before the encounter, the Giotto science working team at Darmstadt met to settle its differences about the distance at which Giotto should pass the nucleus. The spacecraft would go on the sunward side, where the comet was most productive and the nucleus would be well lit. The Pathfinder data showed that, with no correction, Giotto would pass within about 700 kilometres of the nucleus. Who wanted it to go closer? Or farther out?

The preferences of the various principal investigators had not changed, in a debate going back more than a year, but now they had to be reconciled. A constitutional side-issue, as to whether the authority concerning the aim-point rested with the science working team or the European Space Agency, seemed irrelevant. They were all in the same boat.

Keller certainly wanted to get his camera much nearer the nucleus than the Vegas went, but he preferred 1000 or 2000 kilometres to any closer flyby. That would improve Giotto's chances of escaping destruction and allow the camera to see the nucleus from three sides. Studies of the magnetic field and related charged particles would also benefit if the spacecraft survived, because it would allow the same phenomena to be observed on the way out as on the way in.

The scientists with chemical instruments wanted to go as close as possible, to detect short-lived materials near the nucleus, even at the price of losing the spacecraft. Keller pointed out that the camera could not track the nucleus if the flyby distance were less than 500 kilometres. Given a supposed uncertainty of forty kilometres in the comet's position, that meant aiming at least 540 kilometres away. When something of a consensus settled on this, the minimum distance for the camera, as the optimum for Giotto, Keller regretted being so specific.

Roger Bonnet, the director of science, was still in Moscow after the Vega encounters, and he was to telephone Rüdeger Reinhard to settle the final aim-point. This time, Giotto's spirit of mischief took the guise of a deadpan Soviet guard at the Intercosmos front door. He denied Bonnet access to the building where he had his hot-line to Darmstadt.

It was twilight on a Moscow evening, and the Vega teams had dispersed to celebrate or rest after their intensive days with Halley. Bonnet knew that the commands to Giotto for its last manoeuvre were due to go out that night. Mission control was waiting to calculate and check them. The Soviet public telephone system was notorious for its delays. Stranded in the street, the Frenchman was looking at his watch with growing exasperation. Then the secretary of Roald Sagdeev came out of the door on her way home. Bonnet explained his plight and she persuaded the guard to let him in.

On the phone Reinhard summarized the science working team's opinions and told him of the partial consensus. Bonnet heard the most vehement voices favouring the nearer or farther approach to the nucleus. Hans Balsiger spoke up for Dieter Krankowsky's instrument, in which his colleagues from Bern were much involved.

'With the neutral mass spectrometer,' Balsiger told Bonnet, 'if we don't go close enough then we have it on board for nothing.'

So the selected aim-point was 540 kilometres from the nucleus. In the hope of avoiding the fiercest of Halley's dust jets, Giotto should head a little below the line from the comet to the Sun. Combining these instructions with the Pathfinder data, the flight dynamicists at mission control quickly computed the small adjustment to the spacecraft's orbit to achieve the desired flyby distance. Esoc's experts were more familiar than the scientists with the uncertainties both in Pathfinder and in the position of Giotto itself. They therefore played safe, in ensuring that the flyby distance would be, if anything, greater than the stipulated 540 kilometres – certainly not less.

In the small hours of Wednesday 12 March, in response to commands, selected thrusters aboard Giotto burned for thirty-two minutes, and the spacecraft was committed to its flyby path.

Roald Sagdeev and a group of his scientists arrived from Moscow for the Giotto encounter, and Dale borrowed the ESOC director's car to go to Frankfurt airport to pick them up. As the customs officers waved him through, Sagdeev put his arm around Dale and said confidentially:

'David, I'm a very superstitious man.'

'I didn't know that,' said Dale.

Sagdeev produced a small piece of wood. 'This is my lucky charm. Whenever anything went wrong in the Vega project I used to stroke it and it would bring me luck. Now I think Giotto needs it.'

Taking out his pen, Sagdeev wrote on the talisman, 'Good luck to Giotto,' and gave it to Dale.

A cosmic ballet was coming to its final pas de deux. Seen from the viewpoint of the conductor, the Sun, Halley's Comet was hurtling in from the left and moving downwards across the plane of the Earth's orbit. Dust and gas billowed on the sunward side, only to be pushed off into the vast tail pointing away from the Sun. Giotto made its entrance from the right. It was rushing in to meet the comet momentarily centre stage. The comet would kiss the spacecraft as it passed, perhaps fatally. Dead or alive, Giotto would hurtle onwards to exit stage left, while the comet would continue on its path slanting down to the right. The comet's tail filled the stage from front to back, as seen by Giotto during its approach.

The spacecraft named in memory of the first careful painter of Halley's Comet already saw it at much closer quarters than Giotto had in 1301. The glowing tail was 20 million kilometres long and the spacecraft approached it from the side. The streamlined head of the comet, where the target nucleus lay, loomed like a giant light-bulb. If Giotto could have ridden in Giotto, the experience would have been as overwhelming and alarming as a flight into the smoke and fire of an erupting volcano. But even as sharp- eyed an observer as he would have formed only a coarse impression of events. The spacecraft possessed senses far keener than any human's, for the task in hand.

When it was still 8 million kilometres and thirty-two hours away, with the tail spanning a fifth of the sky, the spacecraft began to feel Halley's presence. The smart devices from Holmbury St Mary in England, Bern in Switzerland and Maynooth in Ireland all sensed the subatomic outriders of the comet. These were electrified hydrogen atoms with more energy than the typical sun-blown particles of interplanetary space. The instrument conceived at Toulouse in France joined the comet reporters when it felt electrons going the wrong way, towards the Sun instead of away from it.

As Vega-1 discovered a week earlier, Halley had a bow shock, more than a million kilometres from its core. When Giotto bored through this invisible zone where the comet first hampered the orderly flow of particles in the magnetized wind from the Sun, the sensors saw a flurry of increased subatomic activity. At the same time, the instruments from Cologne in Germany registered an increase in the prevailing magnetism. For scientists hoping to understand the behaviour of electrified matter, the commonest stuff in the Universe, this agreement about the bow shock's position was reassuring.

For its plunge into the head of the comet, Giotto would have to look after itself. The eight-month voyage through interplanetary space had taken the spacecraft 144 million kilometres or eight light-minutes away. This was too far for momentary commands to have any value, especially as Giotto was travelling at nearly 250,000 kilometres an hour in relation to the comet. In the sixteen minutes' it took for data to reach the Earth by radio and instructions to return, Giotto would pass half way through the comet's visible head. The main revelations about the nature of Halley's Comet would come in a period of just four minutes as the spacecraft closed with the nucleus at midnight, zulu time, on 13-14 March 1986.

Six years of effort, and countless hours of overtime in the laboratories and factories of Europe, therefore ended with bursts of final instructions to Giotto from mission control at Darmstadt. Andrew Parkes, as spacecraft operations manager, flew the spacecraft into Halley's Comet. When he was satisfied that it was precisely aligned, bumper shield first, to meet the comet's dust, he cut off the fuel supplies to the thrusters to repress any mad impulse from Giotto itself to alter its attitude in mid-comet. The spacecraft was given the option of switching its radio transmitter to a spare microwave amplifier in case the first should fail. Various housekeeping and data-handling arrangements were 'configured' to their encounter modes. Scientists who wanted their instruments to do different things at various stages of the encounter had to arrange for the commands to go in advance into the onboard memories, with time codes.

The ground station at Carnarvon in Western Australia fired the final pre-planned commands towards the spot in the sky between Sagittarius and Capricorn where Giotto was closing with Halley's Comet. The last went out more than three hours before the closest approach, with 750,000 kilometres still to go. As long as there was no hitch, further uplink traffic was unnecessary.

Six ground stations were standing by. Besides Carnarvon, the European Space Agency had the use of the German station at Weilheim in Bavaria as a second uplink. The Parkes radio telescope, the downlink station in Australia, was already receiving Giotto's signals. The only snag with Parkes was that if the wind blew too strongly the dish would have to stop operating and park in a safe position. The weather forecast was favourable, but it was reassuring that the Deep Space Network station at Canberra was providing a back-up. NASA's other stations, at Madrid in Spain and Goldstone in California, would receive any signals that Giotto might care to send after the encounter, if it survived.

One of the hottest seats at the European Space Operations Centre was occupied by a young German physicist, Gerhard Schwehm. As Rüdeger Reinhard's deputy, he was the mission scientist responsible for liaison between the Giotto experimenters and the flight control team. He had to marshal the requests for sending new instructions to the instruments, decide on priorities, and approve the commands before passing them on to the spacecraft's controllers.

Schwehm knew Giotto and Halley intimately. He came into comet physics as an expert on interplanetary dust particles, and his first association with the spacecraft was as a co-investigator in the two main dust experiments – Jochen Kissel's dust analyser and Tony McDonnell's dust impact detection system. He joined ESOC in Darmstadt in 1983 as a consultant on the vexed question of whether the brightest spot in Halley's heart might not be the nucleus at all, but a concentration of brightly lit dust some distance off. Schwehm found himself interacting with the world's astronomers who plotted the comet's orbit, and he made modern reinterpretations of Halley photographs taken in 1910.

Early in 1985, Schwehm went to Noordwijk to join ESTEC as a planetary scientist, but was at once asked to help Reinhard, who was overworked. At first this involvement with Giotto meant just a few hours a week, but it grew rapidly after the launch. He found himself back in Darmstadt sharing the long, late hours with Howard Nye as the scientists seized the chances to make far more observations during the cruise than originally envisaged.

Schwehm was glad that he had not moved his family to the Netherlands. In the week of the encounter, like many others at Darmstadt, he seldom had more than four hours sleep a day. He would arrive home in the morning just as his children were leaving for the kindergarten, and when they returned at lunch-time he had to wake up and think about going back to mission control.

The evening of the encounter fell on Schwehm's birthday. By way of celebration five edible Giottos, fabricated by a Darmstadt cake-maker, appeared among the consoles at ESOC. There should have been seven cakes, had anyone been able to foresee the future. Before Schwehm's association with Giotto ended, 13 March would come and go at least six more times.

Once the final commands had gone out to Giotto, the scientists had in theory nothing to do except watch the data come in from their instruments. True to form, it was the camera that had problems. As leader of the Giotto camera team for six years, Uwe Keller had known many bad moments, but the worst was when the spacecraft was dashing into Halley at the climax of the mission.

The camera in its full encounter mode began taking pictures three hours before the closest approach. Every four seconds, with each rotation of the spacecraft, a new image was received in Keller's ground equipment. The camera was set to track the brightest spot in the comet, but the rocking mirror that had given problems just before Giotto's launch showed a slight glitch. It omitted a few steps in its rotation, as commanded by the computer program.

As a result the camera system suffered a delusion. It thought that the brightest spot was shifting its direction much faster than it really was, and was therefore overestimating the distance at which Giotto would pass the nucleus. That meant that the camera would look the wrong way at the crucial moments as the comet's nucleus raced past.

In the special room allotted to it in the science area at Darmstadt, the camera team hurriedly analysed the problem and considered ways of curing it. Uncertainty about the true miss distance made this very difficult, and the minutes were racing by. One control engineer in the team was so overcome by the tension that Keller had to take the white-faced man out of the room, to calm him down. About an hour before the closest approach a postscript to the pre-encounter commands went out to Giotto, intended to help the camera retrieve the nucleus if it should lose sight of it.

After he had watched the controllers send off this message, Keller paused to look at his images of the comet. All signals from Giotto's camera were being recorded electronically for later analysis, but some images were also displayed, roughly in real time, for the media and the distinguished guests invited by the European Space Agency to attend the encounter.

At the insistence of a German television producer, who craved striking visuals at any price, the pictures went out to the world in false colours. Each hue was a code for a certain intensity of light, and the TV monitors showed misshapen, garish triangles nested one inside another. In principle, with a coding scale to hand and some minutes' study, an expert might interpret the scene as a fan of brightness in the head of the comet. In practice, the false-colour images mystified everyone that night, including astronomers.

The camera team used the false colours to check that the instrument was gathering light over a wide range of intensities. But to visualize what the comet looked like was difficult even for Keller. Half an hour before the closest approach, when Giotto was still more than 100,000 kilometres from the predicted position of the nucleus, Keller switched his own equipment to a simple, black-and-white mode of operation.

The puzzling triangles gave way to a glorious view of Halley's Comet piercing the black sky with its luminous fountains of dust and gas. There was a bright, point-like centre. Keller peered critically at the image, looking for hints of detail and appraising the sharpness and range of the contrast between light and shade in the centre of the scene.

'We have better images than the Russians,' he murmured.

CHAPTER SEVEN

DUST-UP WITH HALLEY

Although Giotto had become a pure robot, it remained the embodiment of a thousand human minds that with infinite care and a sense of high adventure prepared the spacecraft for precisely this moment. The makers and operators were not alone that night in watching its progress. More than a quarter of the Earth's inhabitants followed the events on television. Robot or not, Giotto had the human species cheering it on.

Its automatic systems were at first preoccupied with gathering observations from the instruments and relaying them to the Earth. Giotto's eye, the television camera from Katlenburg-Lindau, searched ahead of the spacecraft for the nucleus of Halley's Comet. A number of noses sniffed the chemistry of the comet, and one of the instruments from Heidelberg inhaled cometary oxygen an hour after the spacecraft became self-reliant. Before long it was smelling carbon, then broken water molecules, and eventually water vapour itself.

To sense the dust grains that streamed and glowed in the sunlight and made up much of the visible head and tail, the microphones prepared at Canterbury were Giotto's ears. They listened for the drumbeat of dust on the shield, counted grains and judged their masses. The first impacts occurred more than an hour and 290,000 kilometres before the closest approach. The tattoo became faster and faster and nearly all the detectors were hearing hits at 100,000 kilometers.

For sniffing the dust grains and assessing their chemical make-up, Giotto possessed, like the Vegas, another instrument from Heidelberg that registered a patter of tiny grains far smaller than expected when the dust experiments were planned. The sensor took advantage of the violence of the dust impacts, which vaporized the grains, and sorted the ingredients according to their various masses. During the last twenty minutes of the approach, as it sampled many small grains, it found fragments of stony material and savoured compounds of carbon too, of the kinds used by living creatures. These also registered in an instrument from Katlenburg-Lindau, incorporated with the Toulouse experiment.

Even the electronic systems of the spacecraft were severely taxed during the last four minutes and 16,000 kilometres before the closest approach to what was supposedly the Halley nucleus. The sensors worked near to their limits. The drumbeat on Giotto's bumper shield became almost continuous. With 3.5 minutes still to go, the Canterbury dust detectors began registering relatively heavy grains that penetrated the front layer of the shield and pounded on the second.

The sky in Giotto's wake lit up dramatically as the spacecraft entered Halley's visible dust clouds. This registered in the backward-staring optical probe from Verrières-le-Buisson. By the intensity of light at various wavelengths, it recognized various molecular fragments in the atmosphere of the comet and measured the abundances of dust. The rotation of the spacecraft turned a polarizer in the instrument, which added information about the directions of vibration of the incoming light waves.

Giotto passed through the front line of the battle where the cool stuff of the comet contended with the hot and magnetized stuff of the Sun for the command of interplanetary space. Subatomic particles, and heavy molecules too, crowded in a zone of strong magnetism. Then, with just over a minute to go to the closest approach, the energy of the particles plummeted. Their 'kinetic temperature' dropped from the 2000 degrees Celsius of the solar wind to 100 degrees below zero. The Cologne instruments found the magnetism vanishing. Giotto entered a cool world that was pure comet, from which the solar wind was barred, and where no spacecraft, not even the Vegas, had ever been before.

The camera was fixed on the brightest spot in the sky, where the nucleus was supposed to be. It grew rapidly until it filled the field of view with with speckles. Now the camera was supposed to inspect this region from other angles. As Giotto sped forward to pass the presumed nucleus, the camera had to look more and more sideways, in relation to the spacecraft and its track. By nine seconds before the closest approach, the camera's baffle, the tube that screened and aimed the instrument, had swivelled through some forty degrees . . . *Mamma mia!*

Giotto reeled. With 7.6 seconds to go, when the spacecraft was at a slant distance of 790 kilometres from the bright spot, it ran into a jet of dust. One grain, only a tenth of a gram or so in mass but travelling at sixty-eight kilometres a second, hit the spacecraft on a forward corner. As the dust grain splattered on the edge of the bumper shield with the force of an exploding hand-grenade, the off-centre impact set the spacecraft wobbling.

That largish grain and a cloud of smaller grains pounding on the shield slowed Giotto perceptibly. Other grains treated the slanting tube of the camera like the vane of a windmill, and made the spacecraft spin slightly slower. Then they ripped the tube right off. Electrified atoms created by the dust impacts enveloped Giotto in a glow that brought electrical mayhem. It short-circuited components, and high-voltage discharges blasted the delicate electronics of some of Giotto's sensors. The microwave amplifier sending signals to the Earth stopped working.

Until this moment, Giotto's control systems had supervised the gathering of data from the circus of sensors, and relayed them to the Earth. Now the ferocity of Halley's Comet put the spacecraft's capacity for robotic self-assurance to its severest test. If the clever reflexes implanted by the engineers failed to cope with the emergency, even the undamaged sensors would be silenced. The spacecraft would disappear without trace.

Giotto put its first fault right in less than a second, as it switched on the spare microwave amplifier. But almost immediately it lost contact again with the Earth, because of the wobble and the change in the rate of spin. The tilted dish on the back of the spacecraft that beamed the radio signals earthwards required its despin motor to compensate exactly for the spacecraft's rotation. The impact threw out that finely adjusted action so that, at the distance of the Earth, Giotto's narrow radio beam swung drunkenly around in circles millions of kilometres wide. It missed the home planet entirely.

Onboard electronics logged the Sun appearing in a sensor at each revolution of the spacecraft, and discovered that the spacecraft's rate of spin had changed. Until the despin motor corrected its speed, the aiming error would increase by one degree at every revolution. But smooth adjustment to an electric motor took a little time.

Giotto skimmed past the bright spot in the heart of Halley's Comet, a few minutes after midnight, zulu time, on 14 March. The miss-distance was 596 kilometres. By that moment of closest approach, the spacecraft had revolved less than twice since the impact and it was out of touch with the Earth. In all, twenty-two precious seconds elapsed before the system had roughly corrected for the change in the rate of spin.

Even then the wobble made it impossible to keep the radio dish trained faithfully on the Earth. The beam still circled, but now it passed over the planet at intervals of 3.2 seconds, which was the rate of the wobble. The radio telescope at Parkes in Australia registered short bursts of data that reassured the waiting teams that all was not lost.

The wobble was a problem that electronics could not cure, but Dutch engineers had used classical physics to prepare Giotto for such an emergency. Two tubes, each sixty centimetres long and filled with a treacly liquid containing a heavy ball, made nutation dampers. The balls shifted sluggishly as the spacecraft wobbled, in just the manner required to ease the axis of spin slowly back towards the correct line. The last vestige of the wobble would never go away, because the loss of the camera tube and damage to the bumper shield had left the spacecraft unbalanced. But thirty-two minutes after the hit, the sticky liquid had reduced the wobble enough to restore continuous radio contact with the Earth.

By then, with Halley's nucleus 130,000 kilometres behind it, Giotto was heading out of the comet on the far side. It was sensing, in reverse order, changes in particle counts and magnetism similar to those seen on the way in. Even before it had settled its wobble, Giotto found itself bombarded with radio commands that chased it through the head of Halley's Comet. The hours of independence were over. And on the Earth, 144 million kilometres away and receding fast, the fate of the wounded robot was in the balance.

There were always two schools of thought about Giotto's chances of survival in the dust-storms of Halley's Comet. For some scientists and engineers in the project, it was a kamikaze mission that would end with the annihilation of the spacecraft. From much the same information, others judged that Giotto would come through the encounter more or less unharmed.

The spacecraft operations team in mission control took the pessimistic view. From the start of the encounter, Giotto's controllers were waiting for the spacecraft to die. They did not know when it would happen, and hoped it would be later rather than sooner for the sake of the science.

They were quite stoical when the spacecraft's signals flickered and then disappeared, a few seconds before the moment of closest approach. More emphatic than the red danger signs flashing on the display screens was the silence that came with a sudden interruption in the background chatter of printers, which made hard copies of the spacecraft's data transmissions.

The real surprise was the reappearance of telemetry after less than half a minute. It was intermittent, because Giotto was plainly wobbling. Nevertheless, the spacecraft was still alive.

At Tidbinbilla on the outskirts of Canberra, Australia, the NASA deep space station tracked the spacecraft as the back-up to Parkes. The principal investigator for the Giotto radio science experiment was working there. Peter Edenhofer, from the Ruhr University at Bochum in Germany, obtained his insights into Halley from changes in the spacecraft's transmissions. The project management frustrated his idea of measuring free electrons in the comet's atmosphere, by refusing to sanction dual transmissions from Giotto on different frequencies. But Edenhofer expected, and found, small shifts in the apparent frequency of the main transmitter, as collisions with dust and gas slowed the spacecraft down. These Doppler shifts gave clues to the dust's distribution in the head of the comet.

Despite his disappointment concerning the electrons, and his isolation from the rest of the science working team at Darmstadt, Edenhofer was pleased to be taking part in the Giotto mission. He remembered the words of Goethe about the battle of Valmy, *Ihr könnt sagen ihr seid dabei gewesen* – 'You can say you were there.' Less poetically, but with equal warmth, NASA's Australian radio engineers felt the same sense of occasion. When the station saw Giotto disappear and then reappear, one of them slapped Edenhofer on the back and said, 'You've got a tough spacecraft, Peter.'

On the other hand, Rüdeger Reinhard was more shocked than anyone else at Darmstadt when Giotto went off the air, even momentarily. He had lived with the dust hazard since the days of the US-European International Comet Mission. He had supervised the computations of the risk, and the design of Giotto's bumper shield was based on his own reckonings. The statistics suggested that the chance of severe harm from dust grains was only about one in ten. Reinhard had persuaded himself that Giotto would escape intact.

In the outcome both optimists and pessimists were wrong. The loss of signal as a result of a dust-induced wobble had been predicted by Robert Lainé when the spacecraft was designed, but the interruption was later and briefer than he estimated. Giotto survived the encounter, but it was certainly not intact. David Wilkins, the flight operations director, found himself overseeing efforts to assess the damage.

Through the night, while everyone else at Darmstadt seemed to be swigging champagne, the team in mission control had to stay sober and nurse a sick spacecraft. It was the busiest time. Even as the controllers tried to discover the spacecraft's condition, they had to give first priority to the scientists, who were either agitated about injured instruments or anxious to wring the best last data from those that had survived. For six hours Gerhard Schwehm and the controllers were approving and sending the scientists' commands at a rate of one a minute.

Dieter Krankowsky's neutral mass spectrometer was stone dead, as were components of Hans Balsiger's and Henri Rème's experiments. Part of Alan

Johnstone's plasma experiment suddenly expired ninety minutes after the closest approach. Some other instruments were misbehaving. Uwe Keller's camera had switched itself to a safety mode and was sending no images. On the other hand, Fritz Neubauer and Anny-Chantal Levasseur-Regourd, with instruments near the back of the spacecraft, were still receiving data as if nothing had happened. So was Susan McKenna-Lawlor, whose EPONA near the front was protected by a special lip on the bumper shield.

Up till the time of closest approach her whole team was gathered around her, but afterwards nearly everyone melted away, to join in the celebrations that were starting in the science area as well as among the VIPS. McKenna-Lawlor found herself working almost alone, with only a young German technician to help her. Nothing would lure her away while data were coming in. For her, the action was still at Halley's Comet. Around the time of closest approach her energetic-particle telescopes registered a huge surge in the count-rate. While it was partly due to the effect of dust grains impacting on the detectors, the presence of cometary particles in this 'spike' would keep her and the theorists scratching their heads for years to come.

As a matter of public relations, the spacecraft's encounter with Halley's Comet should have been a stunning success for the European Space Agency. Everyone from the media, as well as the agency, wanted a grand night for Europe. The opportunity was fumbled, even though some fifty-six television teams from thirty-seven countries were at Darmstadt for the big event, and their transmissions and relays served a worldwide audience estimated at 1.5 billion, comparable with that for World Cup football.

Many of the viewers were taxpayers of Europe who paid for the mission through their governments' subscriptions to the agency. They were not meanly checking how their money was spent. The celestial apparition fascinated them, and so did the high-tech bid to intercept it. Even if few viewers had a sharp scientific sense of how the scrutiny of Halley's Comet would help in understanding their own existence, they all shared the prehistoric hunch that the key to life's deepest mysteries lay in the heavens.

Europe's television broadcasters tried hard to make it a special event. The BBC had a team at Darmstadt fronted by Patrick Moore, a well-known popularizer of astronomy. It also gathered some top British astronomers at the old Greenwich observatory where Edmond Halley had served as Astronomer Royal. Like other experts, they were baffled by the false-colour images coming from Giotto via ESOC and remained non-committal. The unfortunate presenter at Greenwich was expecting, as a special trick, an immediate display of dust impacts on the spacecraft as registered by the Canterbury experiment. He waited in vain because of a last-minute mismatch of software between the scientists and the broadcasters.

The hours of the encounter, though very short for the scientists, were protracted by the norms of television programming. The BBC concentrated on the science, using pre-recorded visual material and live interviews with the experimenters. Other broadcasters spun out the evening with rock groups and chat shows. In Rome the images from Darmstadt were relayed to a large audiovisual event in the main stadium of EUR. In Paris the brand-new Cité des Sciences et de l'Industrie at La Villette inaugurated itself on *La Nuit de la Comète*. An astrologer solemnly commented on the false colours of Keller's images as if they were real.

There were two main points of interest for the public: what did Halley's Comet look like inside, and how did Giotto survive its encounter? Both were muddled. The false colours of the nucleus seen in the approach gave way to an even more bewildering jumble at the last moment when dust jets largely filled the camera's field of view. The whole world saw or heard the moment when communication with Giotto was lost. That this possibility was always foreseen, and that the spacecraft had already sent back copious data from ten experiments, were simple facts that did not always register with the broadcasters or the public. In Rome the crowd whistled derisively at Giotto's demise.

The lateness of the hour, after 1 a.m. by the clocks of Western Europe, did not help. Many Europeans went to bed believing that Giotto was a disaster. When Anny-Chantal Levasseur-Regourd returned to Paris from Darmstadt, glowing with pleasure at Giotto's success, she was astonished to receive phone calls from friends commiserating with her about the failure of the mission.

For those journalists and VIPS who stayed up patiently at Darmstadt, a little news trickled through. Giotto was back on the air. All the experiments had worked well and some were continuing. Images of the nucleus would be available the next day. Meanwhile, do have some more champagne.

Susan McKenna-Lawlor was conspicuous in a bright red hat that she had been wearing for forty-eight continuous hours with her ground-support equipment. She caught everyone's eye as an unlikely looking principal investigator whenever she emerged briefly from the science area. With her bubbling enthusiasm for the mission and for her own results she won special attention for EPONA, Giotto's smallest experiment. 'It has travelled further and faster than anything else ever made in Ireland,' McKenna-Lawlor said.

When dawn broke, the director general found her and her technician carefully pasting together the paper sheets carrying the wiggling lines of data from her three energetic-particles sensors. At a press conference McKenna-Lawlor was able to ask the projectionist to feed her immensely long chart slowly through the projector, as she said, 'Gentlemen, I invite you to fly through the comet with me.'

Flashes of good presentation came from personal initiative but they often ran unnecessarily late, hours behind the events. Channels of information that worked superbly, from Giotto at Halley via Australia to Darmstadt, broke down at the ceiling between the science area and the reporters on the floor above. Individuals in the science and operations teams found themselves torn between telling the world what was happening and getting on with their work. Howard Nye spoke for many in the Giotto teams when he remarked afterwards: 'If we want to sell this stuff to Europe and the taxpayers then we'd better make it look good.'

Who was to blame? Perhaps no one, unless it was the German television people who insisted on having false-colour images and officials who let them have too big a say in the arrangements. Giotto's most important task was to find the nucleus, but even the best-informed watchers on the Earth could not make it out.

An initially cool response from the media in 1985, after Giotto's launch and before Halley fever had gripped the world, led the European Space Agency to underestimate the interest. By the time it was clear that Darmstadt would be swamped with TV crews, only a few weeks remained to organize the facilities. The Deutsche Bundespost laid on mobile telecommunications systems to supplement the normal facilities of the city of Darmstadt and make sure that ESOC's operations with the spacecraft were never endangered by busy lines. Any thoughts about content and presentation were inevitably belated and disorganized.

Some recriminations afterwards focused on Uwe Keller, as the mission's picture man and the source of those incomprehensible false-colour images. Yet to expect him to shoulder the main public-relations burden would be like asking the leading jockey to commentate on a horse-race. A particular allegation that Keller had excellent pictures of the nucleus and withheld them for devious reasons of his own, was absurd. During the encounter he did not even know whether the camera saw the nucleus of Halley's Comet.

An hour after the closest approach, Uwe Keller despaired of his camera. The beard of the comet physicist from Katlenburg-Lindau had whitened during the years of preparation for this night. In mission control he watched the team try feverishly but in vain to coax the camera to look back at Halley and produce more pictures. With every minute that passed, Giotto left the nucleus 4000 kilometres further behind.

With a shrug, Keller left mission control and walked back to the camera room in the science area. No one could blame him for the damage to an instrument so expensive in cash and anguish. On the contrary, he had pleaded for a greater miss-distance, to reduce the risk of the spacecraft being hit by dust. He noted ruefully that some of the chemical experiments failed to return data from the closest distances demanded in the debate about the aim-point.

The loss of the camera blew away Keller's pre-encounter fear of losing sight of the comet for technical reasons. It also ended his hope of seeing the nucleus from other angles. But up to nine seconds before the closest approach, the camera seemed to have worked well, sending back more than 2000 images.

The main science area at Darmstadt was crowded with members of nine other experimental teams excitedly showing off their results to one another. Keller had to settle down with his own people, examine the images, and decide what they were seeing. Most, like Keller himself, had been too busy with the camera operations, and then with the consequences of the damage, to look carefully at the harvest of pictures. It was time to shut out extraneous thoughts, and the distractions of the Giotto carnival at Darmstadt, and commune with cometary nature.

Fred Whipple, Mr Snowball himself, was present in the camera room as a co-investigator in the experiment. Giuseppe Colombo and Ludwig Biermann, other eminent men who joined Keller's team when he was fighting to save his project, did not live to the end of the mission. Biermann fell ill and died even as Giotto was flying towards Halley. Keller was sorry that his old teacher never saw his generation's pioneering ideas about comets being put to the test by observations on the spot.

Although it was two o'clock in the morning by German time, adrenalin still flowed like the coffee. The team members knew that they had before them the best close-ups of Halley's Comet that anyone had ever seen. Just as psychologists invited people to recognize objects in random inkblots, in the Rorschach test, so the strange pictures demanded an astronomical interpretation. Who knew what a comet's nucleus should really look like?

The word 'snowball' used by Whipple and other comet scientists for many years fostered expectations of an icy object glittering in the sunlight. But the snowball was also supposed to be dirty. After many visits to the Sun's neighbourhood, Halley's surface could be coated with accumulated dust. Whipple and others had even speculated before the encounter that it might have a crust as dark and matt as black velvet. It was a proposition that, like the dark moons of Uranus, left Keller worrying whether his camera would ever see the nucleus amid the glare of Halley's dust. Another possibility, suggested by the Vega images, was that thick dust clouds gathered close to the nucleus would mask any view of the surface, whatever its colour.

Two roundish bright blobs and a dark area beside them figured in all the images as Giotto approached the heart of the comet. Some of those examining the pictures with Keller saw the more prominent of the two blobs as the comet's nucleus. But in that case, what was the dark area in the Giotto images? Whipple was ready with an answer to this question too. He had pondered it while the others were still preoccupied with the operations. With his persistent ingenuity, he suggested that the dark area was the shadow of the nucleus, thrown on to a sunlit dust cloud behind the nucleus. An analogy for Whipple's notion would be the distorted human shadows sometimes seen on the smoke of a nocturnal bonfire.

Even as Keller listened to the arguments, pattern-recognition cells clicked in his brain. He saw the nucleus at last. That is to say, he read the Giotto images of Halley's Comet in a way that he would never have reason to waver about, during the years of enhancing and detailing the data that lay before him. Of all the thoughts and phrases Whipple had come up with, black velvet was right.

'Jesus, this must be the outline!' Keller said.

He ran his finger around the dark area. It was not a shadow but the solid nucleus itself, Halley's core seen in silhouette against a background of diffuse dust. It was very dark in colour and shaped like a peanut shell. The angle of Giotto's approach, in relation to the Sun, gave the nucleus only a narrow sunlit edge, like a thin crescent Moon. But in the closest views there were hints of hills and hollows on the nucleus that cast local shadows on that sunlit edge.

With only a few hesitations, the other members of the team accepted Keller's perception. Even as he spoke, their own reading of the bright blobs changed completely. These were now obviously a pair of intense jets of dust and gas pouring from two main sources on the sunlit side of an otherwise inactive surface. Quick calculations from the known sizes of the fields of view showed that the nucleus was larger than most people had expected. At a press conference, Keller went public with the identification of the nucleus and handed out coloured photoprints in which the dark peanut showed up prominently though fuzzily.

Keller already knew that seeing it had required a dose of good luck, as well as the camera's full sensitivity for detecting slight contrasts in brightness. Had the general dust cloud around the nucleus been thinner, the background luminosity might have been too feeble to silhouette it. If, on the other hand, the dust had been denser, the nucleus could have been lost to view in a sunny fog.

That was what had confounded the camera team of the Soviet Vega missions a few days earlier. Their spacecraft were looking from further to sunward of the comet, whence the intervening dust was thick and luminous, and their cameras were not sensitive enough to show the dark nucleus clearly, beyond the foreground luminosity. It was latent in the signals from the Vega cameras, and data processing would find a dark shape, similar to that seen by Giotto, in time for the first formal publication of results from the missions two months later.

At Keller's moment of recognition in Darmstadt, the essential character of a comet was settled once and for all. The compact nucleus was no longer just a persuasive hypothesis. It was a fact, although 'snowy dirtball' already seemed more apt as a description of the dark nucleus than 'dirty snowball'. Shelf-fulls of speculative books and articles had spun many scientific yarns about comets. Where did they come from? Where did they go in the end? How did they relate to the origin of the Sun and the planets? By breakfast time after Giotto's encounter the basic theories of modern comet science looked stronger.

So did the possible applications to the story of life itself. The sooty colour of the nucleus, especially when combined with the chemical results, could best be explained by the presence of carbon compounds. And if an object the size of Halley's nucleus as seen by Giotto were to hit the Earth, it could easily cause a catastrophe to life like that which wiped out the dinosaurs some 65 million years ago. But new questions were already multiplying that could be answered only by space missions to other comets.

The engineers diagnosed a daunting list of faults in Giotto's housekeeping and control systems, most of which would be impossible to remedy. The small persistent wobble told the flight dynamicists that some unknown mass torn from Giotto had unbalanced it. The star mapper used for checking the spacecraft's attitude, or its orientation in space, was defective. A screen perforated by Halley dust had left the star mapper exposed to sunlight, except on the shaded side of the spacecraft. It could not reliably see the Earth as a reference point.

Important electronic control systems were malfunctioning. Power dumpers that mopped up excess energy from the solar cells seemed to have been damaged. Halley's dust had sandblasted the insulating blankets and white paint, carefully applied to regulate Giotto's temperature, and many parts of the spacecraft were hotter than they should be.

'Switch it off,' David Wilkins advised. 'It's going to be extremely difficult to fly this thing.'

For David Dale the inspection of Giotto's wounds was just a matter of morbid curiosity. As the project manager remembered well, the original plan for the Giotto mission terminated it fifteen minutes after the closest approach to the nucleus. During the cruise to Halley, the scientists asked that ground links with the spacecraft should go on for a day or two longer, to allow for observations on the far side of the comet.

Dale had hesitated even about that, because it was no small matter to arrange. The Earth's rotation meant that the spacecraft would need people watching over it right around the world. Parkes in Australia could cover only part of the added time. The rest had to come from NASA's dishes in Spain and California. Eventually, Dale bargained with NASA to extend the mission for twenty-seven hours after the Halley flyby. When that time was up he would tell Wilkins to let Giotto go.

The European Space Agency's director general invited David Dale for breakfast on the morning after the encounter. Dale had not slept at all, but he imagined it would be just an occasion for more congratulations all round. Instead the director general startled him with the question: 'What are we going to do with Giotto?'

Dale and the agency's accountants reckoned the Giotto mission ended in Halley's Comet. He had bullied and coaxed his teams to the limits of endurance, to make the Halley encounter a success. They solved many tricky design and engineering problems for Europe's first venture into deep space, while working to an inexorable deadline. They rescued the spacecraft from a strike, a fire and a farmer's field. They launched it safely with a slightly dodgy rocket, lost it in space and found it again, and finally guided it to an encounter that historians of science would always remember. Now was the time for everyone to catch up on sleep.

For months past, the project scientist Rüdeger Reinhard had been murmuring about fascinating possibilities for extending the mission if Giotto survived. Roger Bonnet, director of science at the European Space Agency's headquarters, was aware of these ideas. So, it seemed, was the director general.

Dale had stayed resolutely deaf to any such suggestions. As an old hand in space technology, he knew that many a spacecraft survived for longer than intended. Hard decisions often had to be made, to release money and manpower for the endless queue of new missions. Dale's instinct also told him that anything after Halley might be an anticlimax. And now it would be technically very dodgy, according to what Wilkins was telling him.

Euphoria plus hydrazine plus mathematics tipped the balance of the argument against the authoritative opinions of Dale and Wilkins. The success of the Halley encounter made everyone more sentimental than usual about a mere spacecraft. Giotto was bravely cheeping away on the far side of the comet. Why abandon it, when it had plenty of hydrazine fuel left for its thrusters, and when a course of action for preserving it as an operational spacecraft was mathematically beautiful?

Even before Giotto was launched into space in 1985, an orbits specialist at Darmstadt, Martin Hechler, noticed an odd fact about its route to Halley. Unless the comet blew it to bits, Giotto would return of its own accord to the Earth's vicinity in July 1990. By physical law a spacecraft tends to return to its launching place, but the fact that the Earth would be back in position too, and after so few years, was a fluke. No one had given it a moment's thought in planning the Halley interception. A modest correction to the orbit, well within the scope of the spacecraft's thrusters, would make the return to the Earth closer.

'You can bring Giotto back and put it in a museum,' Hechler said.

He was joking. There was no way to slow the spacecraft down for a safe landing. But a close return to the home planet implied possibilities surpassing public exhibition. The Earth's gravity could be used to boot Giotto into a new orbit and send it on its way to another comet. Hechler and his colleagues computed some of the options, and so did Robert Farquhar in the US. The strongest candidate for the new target was Comet Grigg-Skjellerup, which Giotto could reach in July 1992. The scientists pricked up their ears because Grigg-Skjellerup was a faint comet, very different from Halley and smaller even than Comet Giacobini-Zinner visited by the US-led International Cometary Explorer in 1985. The comparisons could be highly instructive.

Hechler updated the Grigg-Skjellerup scheme just two weeks before the Halley encounter. For a flight dynamicist like him, Giotto was a mathematical point tracing a path on a mobile map of the Solar System. Judicious nudges caused the point to coincide momentarily with another moving point that represented Halley, and thence to return to the Earth and fly off to intercept Comet Grigg-Skjellerup. The computer described in moments a voyage through space spanning seven years in all. The orbital mathematics said little about the problems of keeping alive a spacecraft designed for an eight-month mission and severely buffeted by Halley. Nor did it say how an extension might be paid for, out of the overstretched science budget of the European Space Agency.

The director general made up his mind. He told Dale to put Giotto into the orbit that would bring it closer to the Earth. That would at least keep the options open. On the other hand, to look after the spacecraft throughout its four-year return journey was out of the question. Giotto would have to hibernate.

David Wilkins' exhausted team faced another three weeks of activity. It was just as well that the European Space Operations Centre prided itself on responding to emergencies and surprises. While the ground stations needed for talking to the spacecraft were alerted, the flight control team faced the prospect of writing a brand-new mission plan in short order, and setting up the command sequences for sending out Giotto.

First they had to sign off with the scientists. One by one, the experiments in Giotto were switched off. As the allotted twenty-seven hours approached expiry, the science area was almost dark. Hardest to tear away from her monitor was Susan McKenna-Lawlor, the Irish principal investigator. Her instrument named for the Celtic goddess had from the distance of the comet revealed to her, as she said, 'things wonderful and new'. It escaped damage from the dust and it was the very last experiment to close down, at 03.00 zulu on 15 March. EPONA was just then registering an upsurge in energetic-particle activity, and switching it off seemed like terminating a life. Susan wept.

Engineers from British Aerospace had planned to go home to Bristol, to celebrate Giotto's success with their colleagues and families. They found themselves detained at Darmstadt to make a new engineering analysis of the spacecraft they had built. The task was to define a 'survival mode' in which the spacecraft could safely doze unattended for four years and yet remain alert enought for its eventual reactivation.

Big variations in its distance from the Sun during the coming years presented severe difficulties. Giotto would be comfortable only if it sat upright in relation to the plane of its orbit around the Sun. It would then turn in the sunshine like a roast on a spit. With all sides warmed as equitably as possible, vital parts should avoid any extreme overheating or chilling as the orbit carried them nearer or farther from the Sun.

But what was good for Giotto's safety was bad for communications, and for the peace of mind of those who were already looking forward to the awakening and a new mission to Comet Grigg-Skjellerup. In that upright attitude, the radio dish on top of Giotto could not face the Earth. To reactivate the spacecraft after its hibernation, a very powerful radio beam would have to carry the commands, in the hope that Giotto's small auxiliary antenna would receive them. Only NASA's big dishes could shout loudly enough.

Colleagues at NASA's Jet Propulsion Laboratory in California, with unmatched experience of interplanetary missions, made no secret of their opinion of the new European plan. They said it was nuts. But they promised to make the big dishes of the Deep Space Network available again in 1990, when the time came to try to rouse Giotto from its slumber.

The attitude specialists improvised means of finding the spacecraft's orientation with enough accuracy for careful manoeuvres. Although the star mapper could not see the Earth any longer, it could pick out the planet Mars on the shady side of the spacecraft. If nothing else did the trick, the controllers could wave Giotto's radio dish gently to and fro to fix the direction at which its signals were strongest.

Darmstadt's flight dynamicists had to compute precise orbital manoeuvres that would bring Giotto home without actually crashing into the Earth if reactivation failed. 'That would be a bit of a disaster,' Wilkins remarked. The aim was set for a point 22,000 kilometres from the Earth, to be reached on 2 July 1990. The prolonged thruster burns that corrected the orbit to achieve the return began on the night of 19 March, less than a week after the encounter with Halley's Comet.

The burns lasted nearly nine hours in total, spread over three nights. They were tedious for the weary flight control team to oversee, but for the final long burn a table in the control room was decked with a cloth and candles for a 'last supper' of pizza and red wine.

The Carnarvon and Weilheim stations tracked Giotto for eleven days and confirmed that, with only a minute adjustment, the orbit would be correct for the return to the Earth. The final operation at Darmstadt began on 1 April, amid many wisecracks about it being a suitable task for April Fools' Day. Mission control at Darmstadt had to take on trust the last moves towards hibernation. The team instructed Giotto, in advance of the manoeuvre, to slew into its comfortable attitude and then automatically switch off most of its equipment.

The last that Wilkins and his team saw of the spacecraft in the small hours of 2 April 1986 was signals coming in just as it began its turn. They ceased as the radio dish swung away from the Earth. There was no knowing whether Giotto had completed the manoeuvre correctly, and gone to sleep as ordered. In fact Wilkins had no idea how the hibernation and reactivation would work out, because no one had ever done it before.

The Americans had tracked their International Cometary Explorer as it passed Halley's Comet at the relatively enormous distance of 28 million kilometres on

25 March, after flying through Comet Giacobini-Zinner six months earlier at 7800 kilometres from the nucleus. The instruments were working but it was not clear that the energetic particles and magnetic waves that they registered had anything much to do with the distant comet.

The US scientists, never slow to claim a 'first' yet fair with it, contented themselves with saying that the International Cometary Explorer was the first spacecraft to 'investigate' two comets. The Europeans hoped that, just as Giotto went far closer to the nucleus at its first comet, so it might go closer still at its second.

Giotto was asleep on its four-year journey back towards the Earth, the scientists were writing up their results from the Halley encounter, and David Dale was back in Noordwijk catching up on the paperwork. He received a phone call from California. It was the manager of the Deep Space Network, which had given invaluable support throughout the mission. The role of the ground stations in the Pathfinder operation was paid for by NASA itself, but they had helped directly with Giotto on several memorable occasions.

'Dave, I'd like to come to Europe to settle the account.'

'Let's make it Paris,' Dale said. 'I'll give you a nice dinner.' Armed with the Darmstadt log of all the uses made of the Deep Space Network, Dale went to the French capital ready to haggle. But their business was over in a moment.

'Don't you think, Dave, that it's best if we both come in within budget?' the American said. 'Why don't we call it quits?'

Dale was glad on two counts. This bigheartedness augured well for renewed cooperation in Giotto's extended mission, in which NASA's Madrid station would play the same primary role as Parkes in the Halley encounter. And the Giotto project had been saved a lot of money. The scientists were reckoning the success of the mission by their amazing harvests of particles and dust grains. The project manager had worked another kind of miracle by the standards of his own profession. Dale handed back to the science programme of the European Space Agency seven million accounting units of unspent funds.

CHAPTER EIGHT

THE NAKED TRUTH

'Even in science, the journalistic principle applies that first impressions have a validity of their own.' So argued John Maddox, editor of the London journal *Nature*. In an editorial tour de force, he persuaded all the space investigators of Halley's Comet to give him their immediate results. Scientists of many nationalities had to produce short scientific papers in English at a faster pace than they were used to, in a scholarly publication. More compelling than any reasoning about 'first impressions' was the friendly rivalry provoked when Maddox approached the scientific teams of the Japanese and Soviet spacecraft as well as Giotto's. Nobody wanted to be missing from *Nature*'s special issue.

The deadline was just one month after the encounters. Scarcely had Giotto's last experiments been switched off before the principal investigators were in conclave with their teams in various parts of Europe to prepare their instant reports. In Bern, Hans Balsiger had all of his Swiss, German and American co-investigators meeting together in a large room, to confront all the data sent back by the ion mass spectrometer from Halley. The work of analysis was divided between small groups. This concerted effort went on for two weeks. The raw data translated themselves into a vivid chemical portrait of the comet's atmosphere, changing from predominant hydrogen and oxygen far out to predominant water close in, with admixtures of carbon monoxide, sulphur and other materials.

A special hundred-page section of *Nature*, published on 15 May 1986, contained thirty-eight reports on the encounters and the observations of Halley. They proved the care with which the experiments had been prepared, for all the space probes, and the total success of most of them in operation at Halley. As Maddox editorialized: 'Instruments designed in a dozen different countries and launched by three different space agencies have indeed observed the same remarkable object.'

He went on to encapsulate those first impressions. He stressed the huge domain of space contaminated by the comet's emissions, and the apparent confirmation of Whipple's ice-plus-dust theory of the cometary nucleus. He noted the very limited areas of an otherwise coated nucleus that gave vent to jets of dust and gas. And Maddox commented archly on the results from the mass spectrometers:

'When more is known of the constitution of the dust . . . it will be time enough for those who believe that comets carry the raw materials of living things, and perhaps even life itself, to throw their hats in the air.'

Some eyebrows rose higher than the hats, about 'life itself'. For several years past, the astronomers Fred Hoyle and Chandra Wickramasinghe had been claiming that comets were alive with bacteria. In one version of their theory, life in the form of bacteria originated inside the comets. In another, bacteria escaped into space from living planets and hitchhiked on comets to suitable new abodes.

The space missions to Halley found no evidence of live microbes. On the contrary, a dearth of sodium, potassium, phosphorus and oxygen-rich compounds, all ingredients of living cells, undermined the ideas of Hoyle and Wickramasinghe. In that same issue of *Nature*, Jochen Kissel of Heidelberg and his colleagues declared, 'The hypothesis has been disproved at least for Comet Halley as analysed by Vega-1.' But they went on to say, 'We do think that the question of the origin of life in the context of primordial matter as found in comets has become even more exciting.'

Between Vega-1's arrival at Halley and Giotto's exit on the far side, comet science made its largest advance since the days of Halley himself. But the experts would discuss for years to come exactly what they'd learned.

Another deadline loomed. Rüdeger Reinhard, as Giotto's project scientist, was organizing a big international symposium on Halley's Comet at Heidelberg, to be held seven months after the encounter. Astronomers and theorists would be there, as well as space scientists from around the world, including the Americans concerned with the International Halley Watch or the International Cometary Explorer which visited Comet Giacobini-Zinner in 1985. Reinhard convened the symposium on behalf the European Space Agency's space science department. Co-sponsors were the International Astronomical Union and the Committee on Space Research of the International Council of Scientific Unions. If you thought you'd found out anything about Halley, you'd better be there.

As the scientists went on replaying Giotto's signals like a well-worn tune, their minds shifted gear. They left behind their technical and managerial preoccupations with the instruments crammed in the narrow confines of the spacecraft. The sleepless nights counted for nothing, nor did loyalty to one spacecraft rather than another. They had to forget the hothouse atmosphere of the science room at the time of the encounter. The experimenters were back in the wide world of science, with its severe rituals of publication and mutual criticism. All that mattered was whether results were reliable and conclusions convincing.

Concern with finding experimental signals amid the background noise of the instruments and the spacecraft radio therefore gave way to the search for conceptual signals in the noise of copious data. The Nature papers gave only some headline results, and the scientists had to come up with much more thorough analyses, and with carefully referenced comparisons between their work and others'. Results from Giotto had to stand up alongside data from the other Halley probes and the observations by distant telescopes on the Earth and in space.

Not everyone was joyful. At a meeting at the University of Lecce in the heel of Italy in September, the team concerned with Giotto's dust counters, led by Tony McDonnell of Canterbury, ruefully concluded that the most sensitive instrument in the experiment had gone into Halley with a cover still in place. The sensor in question detected the smallest dust grains by the electrical effects of their impact and vaporization. The spring-loaded cover that protected the instrument from deposits during the firing of the Mage motor after Giotto's launch, was commanded to roll back a month before the encounter, but the results showed that it never did so. Only when dust impacts had abraded the cover, by ten seconds before the closest approach, was the sensor fully exposed as intended.

Fortunately for science, another instrument in Giotto was capable of detecting very small dust grains, Jochen Kissel's dust mass spectrometer. This was not functioning perfectly either, as an analyser of the dust's chemical composition, but combining its results with those from McDonnell's dust sensors gave a good count of dust grains of all sizes.

More than 500 experts thronged to Heidelberg at the end of October, for the symposium on 'The Exploration of Halley's Comet'. The scientific papers gathered there would fill nearly a thousand pages of the journal *Astronomy and Astrophysics*. The quantity of results on Halley was not in doubt. But how did they rate in regard to quality?

'A quantum leap in our knowledge of comets,' Asoka Mendis said. This comet scientist from the University of California at San Diego, who participated in two of Giotto's experiments, was awarded the task of summing up the 370 papers of the Heidelberg symposium. He compared what scientists had expected to find with the revelations of their instruments, as evaluated seven months after the encounters with Halley.

Since Ludwig Biermann first deduced the existence of the solar wind from comet tails, theorists had come to visualize several layers where the magnetism and particles around the comet made transitions in mixing with the cometary material. The stages found corresponded broadly with the predictions, but mysteries remained.

Fred Whipple's theory of the 'icy conglomerate' or dirty snowball was the basis of most work on comets. 'The detection of the nucleus of Halley's Comet by the cameras on board the Vega-1 and 2 and Giotto spacecraft came as no surprise,' Mendis said. Water vapour should be abundant in a comet's atmosphere, but ground-based telescopes had never registered it. As Mendis pointed out, the first detection of water vapour in any comet was made by infra-red astronomy from a NASA airborne observatory looking at Halley in December 1985. The space probes duly recorded large quantities of water vapour and confirmed a chemical theory that a comet should contain a charged relative called protonated water or hydronium (H_3O instead of H_2O). Its detection, first at Giacobini-Zinner and then at Halley, was a major finding, because the hydrogen ions added to the water dominated the chemistry of the comet's inner atmosphere.

Water vapour accounted for 80 per cent or more of all the gas released from Halley. The satellite International Ultraviolet Explorer, observing the comet from its orbit around the Earth, measured carbon monoxide as the next most abundant material, at 10 to 15 per cent. Methane expected by many theorists was not conspicuous, and there was a general shortage of carbon in the icy material of the comet. The missing carbon turned up in the dust.

Before the encounters with Halley the belief was that the smallest grains of dust would be about one ten-thousandth of a millimetre wide (a tenth of a micrometre). Grains supposedly of comet dust, collected in the Earth's atmosphere by high-flying aircraft, went down to about that size. Mendis pointed out that the Vegas and Giotto detected large quantities of grains one-tenth as wide, ahead of the comet, despite the supposition that the pressure of sunlight should quickly sweep away any small grains, into the tail.

As to the nature of the dust, stony materials were expected in the form of silicates, similar to those found in meteorites, and there had been speculation about carbon-based chain molecules, and silicate grains coated with carbon compounds. The Vegas provided the first classification of Halley's grains. Some stony grains carried carbon compounds as well, while others did not. The smallest grains, with very little silicate, were named CHONS, because they consisted almost entirely of compounds of carbon, hydrogen, oxygen and nitrogen.

With a volume of 500 cubic kilometres (the size of a large mountain) the Halley's nucleus was much bigger than some prior estimates, but smaller than others. Its irregular shape, variously described as a potato, a peanut and an avocado, surprised no one. The nucleus absorbed 96 per cent of the light falling on it, which, as Mendis remarked, 'puts it among the darkest objects in the Solar System.' Although most experts had expected the nucleus to be much lighter in colour, some had predicted just this degree of darkness.

Mendis invoked a theory, going back to Whipple's early work, that the snow of the snowball could acquire a dark crust of material deposited from the emitted dust. 'What we are observing,' Mendis suggested, 'is not a bare icy nucleus but rather a nucleus covered by a layer of dark, warm dust.' Uwe Keller was going to disagree.

Two elementary questions about Halley remained unanswered. None of the spacecraft said anything about the mass of the nucleus. Attempts to judge the mass, by the effects of the comet's emissions on its motions, left wide uncertainties. Depending on whether the nucleus was honeycombed with empty spaces between fluffy grains, or was more tightly packed, its mass could be anything from 50 billion to 500 billion tonnes.

Unexpected at Heidelberg was a disagreement about the rotation of the nucleus. Exposure of different parts of the nucleus to the Sun, as a comet turned on its axis, would give rise to regular changes in brightness, and some ground observations of Halley in 1910 and 1986 indicated a rotation once every fifty-three hours. The ultraviolet telescope on the Japanese Halley probe Suisei, watching the comet between November 1985 and March 1986, saw its brightness varying over the same fifty-three-hour cycle. But other ground-based astronomers found Halley's brightness changing over a cycle of 7.4 days, and International Ultraviolet Explorer saw variations with the same rhythm.

Confronted with this flat disagreement between respected observers, some scientists tried to figure out if a comet could spin every two days around one axis and once a week around another. The arguments ended in a joke, with a call for papers for a symposium on Halley's Comet after its next apparition in 2061, with the rotation of the nucleus as the first item on the agenda.

While Giotto flew to Halley's Comet, Europe was finding a new identity. Despite several decades of progress towards closer union within the European Community and many other associations, few people had felt like Europeans, as opposed to Germans, Dutchmen, Italians and so on. Another decade or more would be needed to achieve a large internal market with common policies in all areas of common interest. But by 1986, Europe was up and running, especially in science and engineering.

The American post-1945 dominance in fundamental physics was challenged effectively by European accelerators at the CERN laboratory at Geneva, while the European Southern Observatory created the world's largest astronomical facility, in Chile. Names from Europe cropped up more frequently among the Nobel prizewinners, in a confident recovery from the brain drain of the 1930s and 1940s.

Multinational collaborative programmes were becoming the norm in high technology, from microchip development to biotechnology. In the aerospace industry, military and civilian aircraft built from pieces supplied by several countries were offering a serious challenge to American dominance. Europe's Ariane rockets began launching American satellites, just as the French predicted. The European Space Agency benefited from these trends, and plans were laid for an expansion of its work, including more manned spaceflight. Austria and Norway were due to join the agency in 1987, with Finland becoming an associate member.

A general feature of the new Europe, undervalued by those with short memories, was the spirit of peace. While the militarization of space by the superpowers revived fears of aggressive nuclear strikes, and studies of 'nuclear winter' revealed the dangers even to noncombatant nations, there was comfort in the fact that the European Space Agency was confined by its convention to exclusively nonmilitary activities.

Hard on the heels of the Heidelberg meeting on Halley came a gathering in Padua, in November 1986, of the Inter-Agency Consultative Group which had coordinated the Pathfinder operation. After mutual congratulations on the success of everyone's missions and the international cooperation they engendered, the group decided to stay in being as a link between the space agencies of Europe, Japan, the Soviet Union and the US. The study of the solar wind and its effects on the Earth was to be the next focus for coordinated research, with no fewer than twelve missions in the planning stage with the various agencies.

The European Space Agency prepared a small book commemorating the multinational encounters with Halley. Rüdeger Reinhard wrote most of it, and there were lavish illustrations from all the missions captioned in English, French, Japanese and Russian. Pope John Paul II received a copy of the book, at a solemn audience for the inter-agency group in the Sala Regia of the Vatican. Nineteen cardinals were in attendance as the director general of the European Space Agency described to the Pope the collaboration that surrounded the fleet of spacecraft that went to Halley's Comet. He explained how the space agencies and the international space science community strove together to ensure that this once-in-a-lifetime opportunity would not fail for lack of mutual trust. 'The result was the largest space campaign ever undertaken,' the director general commented.

In his reply the Pope said: 'I hope and pray that all of the scientists and engineers in your space agencies will continue to work together in your explorations, and thus merit being called *peacemakers* in addition to your other worthy titles.'

Giotto people had other friendly meetings with heads of state, prime ministers and the like. But as 1986 drew to a close and Halley headed for the outer darkness, comet raiding was losing its glamour. There were plenty of other things to think about: the Irangate scandal in the US; continuing anger between the superpowers; the return to Moscow of an exiled dissident, the physicist Andrei Sakharov.

For dedicated comet scientists in Europe, the quest for further space missions concentrated on a US-German plan called CRAF, which revived the concept of a prolonged rendezvous with a comet, and the European Space Agency's Rosetta scheme for bringing samples from a comet back to the Earth. The first scientific workshop on Rosetta took place at Canterbury in July 1986, and discussions were in progress with the Americans about their necessary participation in the mission.

Among space scientists in general, a burst of excitement about comets was fading. Those of the Giotto experimenters whose interests were in physical questions that could be studied in other parts of the Solar System went in search of fresh missions. Susan McKenna-Lawlor, for one, was planning an instrument for a Soviet spacecraft to study the energizing of particles close to the planet Mars.

The year of the comet had nevertheless galvanized and enlarged the previously tiny band of scientists enthusiastic about comets. And a full decoding of the data required sustained work by a minority of the original scientific teams that sent instruments to Halley. This was not just a matter of filling in details beyond the first impressions, but of defining, even redefining, the character of comets.

So rich in cosmic history was the gas and dust belching from Halley's Comet, to analyse it correctly was a weighty responsibility. The labs were still busy with their task when Giotto reached its second comet, six years later. Fearsome ambiguities delayed the final verdicts on the comet's chemistry.

For Johannes Geiss and Hugo Fechtig, who had pressed for a European mission back in the 1970s, a comet represented a crucial link in the story of our origins. Its chemistry was likely to be nearer to that of the mother cloud of interstellar gas and dust that gave birth to the Solar System, than to the much modified composition of the completed Sun and Earth.

Any material of the mother cloud that collapsed into the Sun itself broke into its elements in that great incinerator. In the inner solar system, only heavy, stony material survived the heat and fierce winds of the young Sun, to be incorporated into the Earth and its rocky neighbours. Within the Earth, heat from collisions and radioactivity, combined with intense pressures, reworked the stony stuff and drained molten iron from the rocks, into the Earth's core.

Comets formed much farther from the Sun and remained small and cool, so preserving their interstellar ingredients. Although seemingly alien in its character, cometary chemistry was not at all remote from terrestrial interests. The loose materials at the Earth's surface, on which life depended, were more like comet stuff than the planet's interior. In the early phase of the Earth's existence, collisions with comets were commonplace events. Comets might therefore be directly implicated in the creation of the ocean and atmosphere, and in supplying the Earth with carbon compounds needed for life to begin.

Giotto went deepest into the comet and Vega-1 gave the best results on the composition of the dust grains. The main instruments and the principal investigators were:

NMS: neutral mass spectrometer (Krankowsky, Heidelberg) in Giotto
IMS: ion mass spectrometer (Balsiger, Bern) in Giotto
PICCA: ion energy spectrometer (Korth, Katlenburg-Lindau) in Giotto
PIA: dust mass spectrometer (Kissel, Heidelberg) in Giotto
PUMA: dust mass spectrometer (Kissel, Heidelberg) in Vega

The analyses were to a large extent a collaborative task for Bern, Heidelberg and Katlenburg-Lindau, with much cross-representation in the experimental teams. Peter Eberhardt of Bern played a prominent part in NMS, and one of the twin instruments of the IMS (HIS) came from Helmut Rosenbauer of Katlenburg-Lindau. The HERS experiment was Marcia Neugebauer's of the Jet Propulsion Laboratory in California.

As the mass spectrometers sampled the gas, the charged particles and the dust grains in the head of the comet, they weighed individual atoms and molecules of various species. Each was roughly equal in mass to some whole number of hydrogen atoms. Data came in the form of numbers: so many counts at atomic mass 18, so many at 71, and so on. Even experts referred to the numbers as 'masses', although strictly speaking each number was the mass of an atom or molecule divided by its electric charge.

The snag was that different materials weighed in with the same mass number. Water, ammonium and heavy oxygen would all show a mass/charge ratio of 18, and so would a triply charged atom of titanium. A fragment with a mass of 28 might be carbon monoxide or ethylene, or hydrogen cyanide that had captured a charged hydrogen atom, or proton. The space chemists concluded that most of the counts at mass 28 were in fact of that last kind.

To crack the comet's chemical code they had to visualize changes that occurred in the head of the comet. If one chemical species was correctly identified, other species would be derived from it, with their own mass numbers. Provided these showed up too, in the signals from the spacecraft, the identification might be confirmed. In the inner regions, newly emitted materials would closely resemble the comet's true composition. As they spread into the outer regions, they would show increasing effects of damage by the Sun's rays and alteration by chemical reactions. Gases released from the dust grains complicated the picture.

The abundant water molecules in the comet's atmosphere tended to destroy many other materials. Attachment of a proton was another common reaction. First, the water molecules would combine with protons, to make charged hydronium, H_3O, and then they would hand over the protons to other molecules that they encountered. As such protonation added one unit to each mass it could threaten to confuse the identifications.

Anny-Chantal Levasseur-Regourd's optical probe in Giotto, infra-red sensors in the Vegas and ultraviolet instruments in the Japanese spacecraft all gave chemical evidence of a complementary kind, in the form of characteristic emissions of light from some of the commoner materials. So did telescopes on the Earth and above the atmosphere, where rockets and satellites could observe the telltale patterns in the infra-red and ultraviolet light from Halley and other comets.

Strong signals at masses 31 and 33 were an example of the codebreaking effort. IMS registered them as Giotto neared Halley's nucleus. Not until 1991 was the team at Bern satisfied that they were due mainly to protonated formaldehyde (mass 30+1) and protonated methanol (mass 32+1). The verdict followed careful modelling by computer of all conceivable physical and chemical processes affecting materials in the range 25 to 35 mass units, as well as many discussions with co-investigators at Katlenburg-Lindau and Pasadena.

That formaldehyde, a compound of carbon, hydrogen and oxygen, was an ingredient of the comet's atmosphere was already indicated by NMS. An infra-red sensor in the Soviet Vega-1 spacecraft also detected wavelengths characteristic of formaldehyde. But the formaldehyde might have come from the breakup of a chain-like polymer of formaldehyde carried in the comet's nucleus and dust grains. Evidence for this polymer came from PICCA. The IMS data implied that new formaldehyde was being released from dust grains at large distances from the comet.

The methanol was an independent discovery of IMS. Also known as methyl alcohol, this was another compound of carbon, hydrogen and oxygen. The scientists had to consider the possibility that the signal at 33 mass units was due to a quite different compound, namely protonated hydrazine, made of nitrogen and hydrogen. They could rely only on general reasoning about the comet's constituents, which made hydrazine an unlikely material, but their bet on methanol was justified when French radio astronomers detected it in another comet.

Discrepancies between the many observations made complete agreement elusive. But as the space chemists savoured the outpourings from Halley, mass by mass, ingredient by ingredient, they formed some strong impressions. NMS indicated a total rate of emission of gases from Halley of about twenty tonnes per second, at the time of the encounter. For comparison, dust appeared to be coming out at less than ten tonnes per second. Within the limits of accuracy of the experiments, you could only say that the dust and gas were of the same order of magnitude.

Of the gases that registered as neutral molecules in NMS and as charged molecules in IMS, most came from primeval water-ice and volatile materials trapped within it. Of the carbon monoxide, accounting in all for about 10 per cent of the gas, half seemed to emerge from the ice and half from dust grains on their way out. Three other volatile ingredients, each at a level of (very roughly) 2 per cent, were the common gases carbon dioxide, methane and nitrogen. The more complex formaldehyde was as abundant, at about the same 2 per cent level – a remarkable result for cosmic chemistry. Ammonia, another common gas previously expected to be a notable ingredient of a comet, seemed rarer. Hydrogen cyanide and hydrogen sulphide were present in trace amounts.

The counts of methane and ammonia in Halley were lower than expected. As the simplest stable combinations of carbon and nitrogen with hydrogen, they would normally form readily in a hydrogen-rich environment. Instead carbon survived in combinations with oxygen that would react easily with hydrogen. The Halley data told of materials kept in deep-freeze in space, in such gentle conditions that chemical reactions occurred very slowly. Unstable combinations of atoms and mixtures of molecules could have endured in the comet since the birth of the Solar System.

Materials present in the dust confirmed this impression. Mineral grains, composed mainly of magnesium, silicon, iron and oxygen, were coated with carbon-rich compounds containing hydrogen, oxygen and nitrogen. In the case of the very small dust grains called CHONS, the carbon-rich material predominated. Even after breakdown, many of the molecules were large and complicated, with high mass numbers. The instruments PICCA, PIA and PUMA were best adapted to their analysis.

PICCA, which Axel Korth flew in Giotto as an addition to Henri Rème's plasma analyser, gauged the molecules present as free-ranging charged particles after dust escaping from the comet decomposed naturally. Korth hoped to measure atomic masses up to 210, but the fierce conditions encountered near the comet's heart limited his analysis to mass 70. He was still able to identify several significant carbon compounds. Besides formaldehyde, these included acetaldehyde and acetonitryl, familiar enough to chemists, but other products suggested the presence of 'parental' materials unlike the ordinary carbon compounds on the Earth.

An absence of some masses deleted at a stroke an entire class of carbon compounds from Halley's major ingredients. If the comet owned hydrocarbons of the kinds familiar on the Earth as fuel oils, many fragments with specifiable mass numbers should have shown up in Korth's data. They didn't, although those hydrocarbons would have been commonplace end-products of many chemical reactions. Instead, PICCA detected 'unsaturated' hydrocarbons with a shortage of hydrogen atoms that left them chemically hungry.

Jochen Kissel's instruments, PIA in Giotto and PUMA in the Vegas, sorted the molecular fragments forcibly set free when dust grains from the comet slammed into a target. In PIA, the target was a tape of platinum foil doped with silver, which spooled past an aperture to keep the target fresh as the impacts multiplied. PUMA was almost identical, the only significant difference being the use of pure silver targets, one on a spool (Vega-2) and the other fixed and corrugated (Vega-1).

PIA and PUMA were easy to spot on the sides of their respective spacecraft because of the distinctive V shape formed by electric racetracks for molecular fragments. As each dust grain hit the target, the shock threw some fragments off the back of the grain, while the heat of the impact vaporized the rest. Much of the material became electrically charged. The equipment accelerated the charged fragments into the racetrack, where the time that each fragment took to reach a detector at the far end was a measure of its mass. The heaviest particles arrived last. Kissel had exciting results from PUMA before Giotto reached Halley. The Vega-1 data were especially good, which was lucky, because the breakdown of an amplifier in PIA limited its analysis to the most abundant materials.

In the months and years that followed, Kissel could compare his results from Vega and Giotto with those from the other chemical instruments in Giotto, notably PICCA. His closest partner was a chemical expert, Franz Krueger, who lived in Darmstadt. Permutations of molecules that might match the observed masses became increasingly numerous and daunting at the higher masses. Informed guesswork was unavoidable.

'A badly adjusted diesel engine would give much the same results as Halley's Comet,' Kissel commented wryly.

Consistency was a key test, for the higher masses as for the lower ones. For example, for the PUMA counts at mass 78, a molecule with five carbon atoms, one nitrogen and four hydrogens would give the right mass. To a chemist this denoted a ring-shaped compound called pyridine stripped of one hydrogen atom. Pyridine was an ill-tasting material often added to industrial alcohol to make it undrinkable, but to biochemists it was a unit in many molecules of life. In the violent circumstances at Halley, certain fragments of pyridine were bound to appear as well. One was acetonitryl, consisting of two carbons, one nitrogen and four hydrogens, giving a mass of 42. Pyridine was the right answer for mass 78 only if corresponding counts for acetonitryl showed up at mass 42. Sure enough, they were there.

From such coincidences Kissel and Krueger drew up a list of carbon compounds in the comet grains, ranging from hydrogen cyanide, a very simple molecule, to adenine, known in the nucleic acids of life. And in the Vega data, Kissel and Krueger saw the same chemical hunger as Korth did in his Giotto results.

In the cold of a comet, many carbon compounds survived that were incomplete by comparison with their stable cousins, like the normal hydrocarbons. In the warmer conditions on the Earth such substances never accumulated because they quickly combined with other materials and disappeared. Those in the comet would react avidly too, if given the chance, for instance by freeing them from their supercooled tomb in the comet and plunging them into liquid water. The suggestion that life began that way belongs to a later part of the story.

'Yes, a decade is quite a considerable fraction of the professional career of the persons involved,' Uwe Keller said. 'But that is typical for a space experiment.'

He was speaking of the task of preparing and operating the Halley Multicolour Camera in Giotto, and then analysing its images of the nucleus. Had the camera survived for the Comet Grigg-Skjellerup encounter, the period of work would have been longer still. As it was, Keller and his closest colleagues at the Max-Planck-Institut für Aeronomie at Katlenburg-Lindau did not finish cataloguing the Halley images until 1991.

As the surface of the nucleus absorbed all but 4 per cent of the light falling on it, the details were hard to examine. Even the sunlit parts stood out no more plainly than writing done with charcoal on a blackboard. Yet, after meticulous processing in Keller's lab, the images yielded prime information on the nature of comets.

From more than 2300 images taken with Giotto's camera at Halley, about 2000 went, partly processed, to the Jet Propulsion Laboratory for incorporation in a compact optical disk published for the International Halley Watch. A selection of fifty images, more thoroughly processed, went into the Catalogue that marked the completion of the Halley Multicolour Camera project. Like commentaries on paintings by old masters, explanations pointed out every feature that anyone had spotted, and every generalization about the comet that the images allowed. The names of an inner circle of investigators would appear with Keller's on the Catalogue: Werner Curdt, Rainer Kramm and Nicholas Thomas. Rüdeger Reinhard helped by editing the Catalogue.

Their chief disappointment was that they saw only one side of Halley, and one cross-section. Up to the moment of the last good-quality image, before the dust impacts damaged the spacecraft and disabled the camera, the angle of view changed by only a few degrees. Not only was the opportunity lost to see more features of the surface, but the overall shape and orientation of the nucleus remained undefined.

The Soviet Vegas sent back images of the nucleus from other angles. After years of data-processing, and comparisons with Giotto's images, Hungarian members of the Vega camera team suggested that the nucleus was 15.3 kilometres long. If so, the apparent length of 14.2 kilometres seen from Giotto meant that the nucleus was slightly tilted away from the European spacecraft, as it approached.

Giotto beheld the comet like a crescent moon, one quarter lit, with the dark side silhouetted against a background of illuminated dust. As the camera was programmed to lock on to the brightest spot, it looked ever more intently at a glow beside the northern corner of the nucleus, where the strongest emissions of gas and dust were occurring at the time. The rest of the nucleus was gradually lost beyond the edge of the frame. The last image showing the whole nucleus was taken when Giotto was still 14,000 kilometres from the encounter. The camera could not at that time resolve anything smaller than 320 metres. A final set of useful close-ups, sent when Giotto was about 2000 kilometres from the nucleus, showed details down to fifty metres on the northern sunlit corner.

Keller, Kramm and Thomas combined six processed images obtained at various stages in the encounter to assemble a view of the whole nucleus in which each part was represented as clearly as the Giotto data allowed. The districts farthest from the bright spot had the least detail, and those near to it the most.

'In future' Keller remarked, when he handed a print of the composite image to a visiting journalist, 'you can judge whether your encyclopaedia is any good. Does it have the best picture of Halley's Comet that will be available until 2061?'

'But is this the naked truth?' the journalist asked. 'Or is there an artistic hand at work?' He had treasured as souvenirs the fuzzy images of the Halley nucleus released by Keller at Darmstadt, on the day after the encounter. Seeing the processed images a few years later, he was a little suspicious about the sensational improvement, with hills and hollows plainly visible on the comet's surface.

Keller replied with a sketch of modern image-processing by computer. It required no artistic licence, and relied on no prejudices about what a comet nucleus ought to look like. When he and his team set about cleaning and enhancing the raw signals transmitted from Giotto's camera, they first excluded obvious blemishes. Cosmic rays passing through the camera produced irrelevant spots or streaks of light in half of all the images. Electrical interference produced stripy patterns like those sometimes seen on TV pictures on the Earth.

To reduce random interference, or 'noise', the team took an average of a few images before and after the image under study. As the comet should look the same in all of them, any differences were due to noise, which could be subtracted. Other imperfections in the raw data came from the camera system. Stray light scattered inside the camera produced some false readings. No optical system could focus light perfectly, but images of a star, Altair, observed by Giotto's camera in February 1986, allowed the team to measure the spread of light. Instead of the perfect geometric point that would be the ideal image of a distant star, the light covered a small disk. The team could then use a mathematical formula to 'deconvolve' the images and so sharpen them without cheating.

In the camera's array of charge-coupled devices, the electronic detectors that registered the light, individual elements varied a little. They gave different signals

The nucleus of Halley's Comet, about 15 kilometres long, as seen by Giotto's Halley Multicolour Camera. The Sun (left of picture) illuminates the nucleus like a crescent Moon and provokes outbursts of dust and gas from the surface. This composite of six images, from different distances, gives a resolution of 320 metres in the lower part of the nucleus and 60 metres near the bright area at the top. The image-processing is by Uwe Keller, Rainer Kramm and Nicholas Thomas. (Max-Planck-Institut für Aeronomie, Katlenburg-Lindau.)

for the same intensity of light. The team mapped these variations, detector by detector. Some overheated detectors gave false signals, causing spottiness. As each fault was identified, the computer corrected the intensities element by element.

'Unsharp masking' was the last, wonderful step that revealed delicate features in the Halley nucleus. Adopting a technique often used by astronomers, Keller's team subtracted from the processed image, element by element, a slightly defocused version of the image. This left a grey image in which the brightest areas were darkened, while slight differences in brightness became relatively conspicuous. When the team then 'stretched' those small contrasts across the full range from white to black, Halley's landscapes emerged as if from a fog, complete with their hills and hollows.

Keller was not content to point like a tour guide to the surface features visible in his processed images. While his team made 3-D models of the Halley landscapes by computer, district by district, to match the light and shadows seen in the images, he looked for physical meaning in the features. A crater-like depression, about two kilometres wide and up to 200 metres deep, was conspicuous on the sunlit side as seen by Giotto, and it contained three bright spots. It had no raised rim, so it was not an impact crater. In Keller's opinion, the 'crater' had shed a lot of material into space in the recent past, but was inactive at the time of Giotto's flyby. Its depth would be accounted for, if it sacrificed a six-metre thickness of dust and ice at each of thirty visits to the Sun, since the earliest recorded apparition of Halley's Comet occured in 240 BC.

The processed images clarified the jets of gas and dust emanating from the comet. These were engines by which Halley produced its celestial fireworks-show. Although the space all around the nucleus was bathed in glowing gas and dust, it was brightest on the sunward side, where three active regions were visible on the edge of the nucleus as Giotto approached: large in the north, intermediate in the middle and small in the south. Seventeen filaments streaked outwards from the nucleus, spreading like the sticks of a fan on the sunward side of the comet. Three or four of them traced back to areas about 500 metres wide in the active region near the northern tip of the nucleus. One of them pointed across Giotto's track and Keller suspected it of being the jet of dust that damaged the spacecraft and crippled the camera just before the closest approach.

The emissions were less vigorous in the bright patch in the centre of the sunward side of the nucleus. Detailed analysis showed three distinct bright areas within the patch, and two emergent filaments. The small active region at the southern tip seemed to be associated with faint filaments too, but most of the filaments originated from sources beyond the horizon as seen by Giotto.

Even the strongest filaments represented only a small part of the general release of material from the comet, and visible dust surrounded the nucleus on all sides. Dust experiments in Giotto (those of Tony McDonnell, Jochen Kissel and Anny-Chantal Levasseur-Regourd) showed no sudden increase in the abundances as the spacecraft advanced from the dark side of the nucleus into the region of space directly exposed to emissions from the sunlit side. Keller supposed that breezes near the nucleus carried dust from the active regions to the dark side.

As its name implied, the Halley Multicolour Camera had filters for assessing the intensities in different colour bands. Although the nucleus was almost black, it had a reddish tinge, which was least pronounced in the three bright spots in the 'crater'. The surrounding dust in the comet's atmosphere was redder than the nucleus itself. Ground-based observations of comets suggested that they grew redder the more times they visited the Sun's vicinity. Halley was intermediate in redness between the almost colourless Schwassmann-Wachmann 1, which always remained far from the Sun beyond the orbit of Jupiter, and the very ruddy Comet Neujmin 1 which seemed to be on the verge of extinction.

The first surprise about the nature of Halley's Comet was plain from the moment when Uwe Keller identified the nucleus in the wee hours at Darmstadt, on the night of Giotto's encounter. Most of the surface was inactive. A second cause of wonderment was on a slower fuse, and much image processing went into confirming it. This was the lack of anything on Halley's Comet that looked remotely snow-like.

A fallacy about the Giotto mission, shared even by some experts, was that the spacecraft's pictures endorsed Fred Whipple's theory of the comet nucleus as a dirty snowball. Not for the first time, scientists were reluctant to shift their beliefs by more than the narrowest margin needed to fit new evidence. When Giotto revealed how very dark Halley's nucleus was, Asoka Mendis and others simply wrapped the good old dirty snowball in a coat of dust gathered from the comet's own emissions. But those who spoke as if Giotto's images proved this opinion were indulging in careless talk. The chief analyst of the images thought nothing of the sort.

Keller assessed the small active regions emitting gas and dust and concluded that they were scarcely lighter in tone than the rest of the comet's dark surface. They absorbed at least 93 per cent of the sunlight falling on them, as compared with 96-97 per cent elsewhere. These were certainly not gleaming ice sheets peeping through a superficial layer of dust, which scientists had expected to see in the comet's zones of activity. The more he studied the images, the more sceptical Keller became about the dust-encrusted snowball. The lack of any sign of subcrustal ice was not his only reason. Halley was also craggy.

The highly irregular peanut shape of the Halley nucleus could never, in Keller's opinion, be fashioned from a vaporizing snowball, which should make a smoother and rounder nucleus. A peak about a kilometre high, named the Duck Tail, was silhouetted as a sharp corner at the southern end of the dark edge. The Mountain was another peak of similar height in the middle of the dark area seen by Giotto. Its summit caught the rays of the Sun. What process, Keller asked, could keep such protrusions neatly covered with a thin crust of dust? And if these peaks were fashioned largely of ice, would they not be especially liable to draw heat from the Sun and quickly disappear?

The Soviet Vega spacecraft measured the temperature of the sunlit side of the surface of Halley's nucleus. The answer was about 125 degrees Celsius, well above the boiling point of water even at the pressure on the Earth's surface. Long-exposed parts of the comet surface could therefore possess no ice. But the active regions were belching steam as well as dust, and there was no sign of any fissure or volcano-like vent to bring the jets from the comet's interior, nor any mechanism to raise steam below the surface. So the active regions had to retain their own ice at the surface. Yet they looked almost the same as the ice-free, inactive regions.

The evidence of his images forced Keller to change the recipe for a comet. Fred Whipple had correctly visualized the ingredients of the nucleus as ice and dust. He also considered the possibility that the dust was more basic than the ice, when he wrote in 1950, 'A model comet nucleus consists of a matrix of meteoritic material with little structural strength, mixed together with frozen gases – a true conglomerate.' But a postscript to Whipple's early description encouraged a belief that the ice was the glue that held the nucleus together:

'If the primitive ices constitute a large percentage of the total mass, the comet truly disintegrates with time.'

In keeping with this latter idea, the dirty snowball captured popular attention and erudite imagination. By 1988, with Giotto's enhanced pictures in front of him, Keller was reverting to something closer to Whipple's first thought. Keller described the Halley nucleus thus:

'The surface of the nucleus appeared uniform. Similarly, there was little variation within its active areas, implying that the interior and the surface were of the same physical quality. No icy surface was visible. The rather large topographic features, in particular the height of the Mountain, eliminate the picture of a shrinking ice ball covered by a regolith of larger dust particles. This all supports the picture of a nucleus the physical structure of which is dominated by the matrix of the non-volatile dust rather than by the volatile material (ice).'

A homely analogy made the distinction plainer. Exposure to the Sun, in the vacuum of space, freeze-dried the material of a comet. If the dirty snowball model were correct, a handful of comet-stuff would be like a chocolate sorbet. When the ice vaporized, one would be left with loose chocolate powder. But in the snowy dirtball that Keller saw, the comet-stuff was like chocolate cake just out of the freezer. It too incorporated frozen moisture. But when that vaporized, what would remain? A lump of cake – Keller's 'non-volatile matrix'. Although freeze-dried and very crumbly, perhaps with pieces missing, the cake would still be recognizable.

For the recipe for the cometary cake, Keller went time-travelling into interstellar space 4600 million years ago. Then, a cloud of gas and dust was gathering to give birth to the Sun and its planets. Fluffy dust grains with both stony and carbon-chemical compositions, and perhaps rimed with ice, were loose in space.

As the density of the mother cloud increased, at the dawn of the Solar System, the grains clumped together into a dark and porous bulk material. The ice content in the pores may have increased rapidly then. In the course of perhaps a million years, the material gathered into comets. These were assembled, Keller suggested, from building blocks of comet-stuff 500-1000 metres wide.

The Giotto images showed several features of that size in Halley. Apart from the Duck Tail and the Mountain, the Chain of Hills north of the crater included four mounds about 500 metres wide and 800-1000 metres apart. This typical scale, Keller believed, was a relic of the comet's formation from the coalescence of thousands of primordial blocks, which came together during the planet-building era. If that was so, the assembly must have occurred a long way out from the Sun, where orbital speeds were low. Otherwise the fragile blocks would smash one another. Even slow collisions would crush parts of the blocks, producing relatively dense zones inside the comet, but the very weak gravity could also leave large hollows between the component blocks.

From the abundance of material in the meteor streams that Halley left smeared along its orbit, comet scientists supposed that the nucleus was much larger when it first entered its familiar orbit, perhaps hundreds of thousands of years ago. So large an object would be very likely to break up, as many comets had been seen to do. Keller wondered if the strangely straight edge of the dark side of the nucleus, as seen during Giotto's flyby, might mark a fracture. And he thought that the irregular shape of the rest of Halley's nucleus preserved a memory of original blocks that were buried together at random when the comet was formed, and became exposed as overlying pieces broke away.

The precise recipe for a comet was no mere technicality. Apart from the questions about the origin and behaviour of comets, it bore strongly on their ultimate fate. Some comet scientists continued to assert, in line with Whipple's suggestion of 1950, that comets simply disintegrated when their ice was exhausted. But the invisibility of the ice, even in Halley's Comet in its prime, made this point of view harder to sustain.

Keller's reading of Giotto's pictures accorded to a comet a structural integrity not dependent on its cargo of ice. It lent support to the notion that many dark asteroids orbiting near the Earth, and threatening to collide with it some day, were old comets that had literally run out of steam.

CHAPTER NINE

SWINGING BY THE EARTH

GRAVITY gave to every orbiting object a memory of where it came from, and Giotto kept revisiting the spot in the Solar System where the Earth spawned it in July 1985. Five times after its rough ride through Halley's Comet, the spacecraft's gravitational instinct brought it back to its birthplace at the Earth's position for July – but at the wrong month. Giotto took ten months to orbit around the Sun, as against the Earth's twelve months. After six orbits Giotto would gain a whole lap around the Sun, and so find the Earth at home in July 1990.

That was the key to extending its mission to a second comet, yet hardly anyone gave a thought to the spacecraft when it began its long, silent cruise in hibernation. Giotto was a penniless waif in the European Space Agency's budget. Engineers and flight controllers had other projects to keep them occupied, while the scientists were either mulling over the Halley results or looking for new missions.

One exception was Peter Edenhofer, from the Ruhr University of Bochum, who was the principal investigator for radio science. He noticed that the spacecraft would be half a lap ahead of the Earth at New Year in 1988, and would therefore lie on the far side of the Sun as seen from the Earth. If Giotto's transmitters could be switched on for a period straddling that date, measurable effects of the solar atmosphere on the radio signals passing through it would give unprecedented insight into free electrons in the Sun's atmosphere.

The agency's science advisers thought it a beautiful idea, but the project management and operations experts ruled it out. Trying to reactivate Giotto at its maximum range would be risky even if the Sun were not interfering with the signals. And a necessary ground station in NASA's Deep Space Network would be closed for upgrading during the period of interest, when Giotto passed behind the Sun.

Edenhofer's proposal nevertheless prompted a few people to think about

Giotto's eventual reactivation. A question for Rüdeger Reinhard and his scientists was what the target should be. Everyone was learning to spell Grigg-Skjellerup, as that comet was the most likely option, but the scientists gave careful consideration to two others reachable by Giotto in the early 1990s. Comet du Toit-Hartley was somewhat too far from the Sun, and Giotto would be cold and underpowered during an encounter. For Comet Hartley 2, another interesting candidate, uncertainty about its movements made a reliable encounter difficult to stage-manage.

Grigg-Skjellerup made one of its five-yearly visits to the Sun in June 1987, so the European Space Agency asked the world's astronomers to track it, with a view to predicting its next apparition accurately. Those at the European Space Operations Centre at Darmstadt who would have to accomplish the feat of reactivation made a start on defining what came to be called the Giotto Extended Mission.

The choice of Grigg-Skjellerup fixed the overall timetable. The spacecraft had to be reactivated in the winter of 1989-90, and lined up accurately for the swingby at the Earth in July 1990 that would change its orbit. If all went well, Giotto would begin a two-year cruise to its second comet. It would be put back to sleep for most of that period, and reactivated again in time for the encounter with Grigg-Skjellerup in July 1992.

At Darmstadt, Martin Hechler, who first proposed the target, teamed up with another of ESOC's experts on spacecraft orbits, Trevor Morley, to study the second encounter with a comet. Distinguished by his big eyebrows and a ready laugh, Morley was to take charge of Giotto's route in space. He calculated Grigg-Skjellerup's motions from the best available astronomical data. To suit Giotto and its handlers as well as possible, the flyby could be adjusted a little as to timing. This could vary important bearings during the encounter, especially those of the Earth and the Sun, which would determine the angle of attack – the orientation of the spacecraft on its track through the comet. Nothing was ideal, but an encounter on 10 July 1992 would leave important options open.

In 1988 the science programme committee at the agency's Paris headquarters authorized Giotto's reactivation in 1990. The delegates postponed a decision on the full Giotto Extended Mission to Comet Grigg-Skjellerup in 1992 until they knew what state the spacecraft and its instruments were in. Everyone was asking whether Uwe Keller's camera could be revived after its blinding by the Halley dust-storms.

Protagonists of the Halley mission had dropped out of the project. David Dale was overseeing the mission management for all scientific spacecraft. After coordinating the science for solar-terrestrial physics missions, Reinhard was becoming involved with plans for an International Space Year to be held in 1992. Giotto's new project scientist was Gerhard Schwehm who, as Reinhard's deputy at the climax of the Halley mission, had liaised with the flight operations team and shared his birthday cakes with them.

The project manager for the Giotto Extended Mission was Manfred Grensemann, a chubby, pipesmoking German with a private enthusiasm for old aeroplanes. His experience with Europe's scientific missions went back to the ESRO-1 satellite of the 1960s. He was deputy project manager for Geos, the spacecraft on which the concept for Giotto was first based. Administrative work in the science department during Giotto's Halley phase left Grensemann fully aware of the debates about extending the mission, and the problem of funding it out of the European Space Agency's small change.

If Giotto to Halley was a cut-price venture, the extended mission was in the bargain basement. NASA bartered support from its Deep Space Network of ground stations, in exchange for European help with other missions. At Darmstadt the orchestra that the flight operations director, David Wilkins, would conduct at mission control was even smaller than before. Howard Nye, now designated spacecraft operations manager, would have about nine other operations staff. Morley and a few other flight dynamicists would complete the team. Engineers from some of the companies that built the spacecraft would be in attendance. The Giotto electronic simulator was dusted down and revived.

Everything depended on the first reactivation of Giotto during its approach to the Earth, and that promised to be ticklish. With British understatement Nye said, 'It won't simply be a case of commanding the transmitter back on.'

The operations team had to solve the puzzle that they set themselves on April Fools' Day 1986, after the Halley encounter, when they hurried to make Giotto comfortable for its hibernation. They had no option but to tilt the radio dish away from the Earth, leaving Giotto deaf. The only radio receiver still switched on was connected to the small look-anywhere antenna.

By the time Giotto was within range of the European Space Agency's small ground stations, it would be far too close for checkouts and manoeuvres before the Earth swingby. Only the powerful transmitters of NASA's Deep Space Network could rouse Giotto from its slumbers in good time. An American deep space probe, Galileo, would keep the NASA dishes occupied until February 1990. Then they would allot to Giotto a period of thirty-five days for the reactivation.

The flight controllers would have to slew the spacecraft to bring it back into full communication. A journalist likened their task to telling a small child by telephone how to unwrap a parcel. The openable end was the one carrying Giotto's radio dish, which had to face towards the Earth, with an accuracy of one degree, before any two-way communication could begin. But which way up would the parcel be, when the gravitational postman delivered it? What would Giotto's attitude be, its orientation in space?

The team knew what attitude they ordered Giotto to adopt for its hibernation. There was no telling how well the spacecraft carried out this manoeuvre. And over four years the spin axis would have swayed. At best the flight dynamicists could guess the tilt of the spacecraft's spin axis to within about forty degrees.

Giotto in its prime, before its injuries at Halley, had clever sensors and automatic controls designed to help it locate the Earth. The team had to assume that these were unserviceable, and perhaps even unsafe to use. Giotto was built for a working life of eight months and had spent fifty-four months in space already. These included periods of overheating when its orbit took the spacecraft relatively close to the Sun, with its thermal control system badly damaged by Halley dust impacts.

Engineers from British Aerospace, Dornier and Fokker studied the defects in onboard systems identified after the Halley encounter, together with the possibilities of subsequent failures or resettings within the spacecraft. Hope lay in Giotto's backup systems, but the correct commands to send to Giotto depended on what was working and what was not. The flight control team drafted a long series of alternative command cycles, to deal logically with possible conditions of the spacecraft.

Commands to Giotto's thrusters would slew the spacecraft about, in an attempt to sweep the beam of its radio dish across the Earth. To avoid putting it into a completely unknown attitude, each command cycle was supposed to end by bringing the spacecraft to its starting point, but every blind manoeuvre was bound to add to the uncertainty about the attitude.

By February 1990, Giotto was chasing after the Earth on its inside track. Communications difficulties with the Galileo spacecraft, bound for Jupiter, had put the NASA ground stations under severe stress. Little more than a week after Galileo had swung by Venus, on its way to Jupiter, the teams and stations of the Deep Space Network reported themselves ready to help the Europeans with their comet probe.

Although the chief NASA station for the Giotto reactivation was called 'Madrid', it stood fifty kilometres west of the Spanish capital, beneath the Guadarrama mountains and El Escorial, the 16th-Century monastic palace of Philip II. At midday on Monday 19 February, the station's 70-metre radio dish was facing eastwards, over the city of Madrid, along the Mediterranean Sea, and up towards Giotto's computed position in the sky. As the Earth turned, and the spot rose higher above the horizon, the dish slowly tilted upwards. Thanks to the law of gravity, the surest part of the whole operation was the prediction of the spacecraft's orbit.

Giotto had been silent for forty-six months. Mission control at Darmstadt began the attempt to revive the hibernated spacecraft at 12.45 zulu, or 1.45 p.m. Darmstadt time. Commands went out to the headquarters of the Deep Space Network in Pasadena, California. They came back across the Atlantic to the Madrid station. Its hundred-kilowatt radio transmitter fired the commands into the big dish, which focused them into a beam, so that an electromagnetic shout rang across 102 million kilometres of space.

'Giotto, wake up!'

The first commands, about 150 in all, ordered the spacecraft to switch on internal power supplies and other pieces of equipment. For more than two hours, while these preparatory instructions were going out, the team at the European Space Operations Centre had no way of telling whether Giotto was listening to them. At 14.55 a sequence of commands told Giotto to start transmitting on its small antenna.

Compared to the hundred-kilowatt beam from Madrid, the unfocused broadcast from the five-watt secondary transmitter in the spacecraft was like a glowworm to a searchlight. The ground station had to switch from being an intense transmitter to a very sensitive receiver. The signals travelling outwards at the speed of light took nearly six minutes to reach Giotto. Any response would take as long to reach the Earth. After sending the transmit command, the controllers sat back.

They waited for eleven minutes and twenty-seven seconds. Then a faint radio wave bounced gently on the 70-metre dish at Madrid. Its source lay less than one-hundredth of a degree from the computed position of the comet probe in the sky. Giotto was alive, awake, and humming softly in the hot sunshine.

That was good news, but the difficult part of the operation was still to begin. The transmissions were too feeble to carry any information in the form of telemetry data from the onboard systems. While the Darmstadt team was checking its command sequences for the following day, Pasadena came through with an unexpected word: 'The spin rate of the spacecraft is 15.31 rpm.'

NASA's experts, eavesdropping on the signals, achieved a technical tour de force in measuring the spin rate. They analysed and reanalysed Giotto's feeble hum, and in just a few hours' work they detected a barely perceptible yet rhythmic change in radio frequency as Giotto rotated. The antenna, perched on its tripod, was not exactly on the spin axis of the spacecraft, since the loss of part of its structure in Halley left Giotto slightly unbalanced. As a result, it swivelled around the axis in a small circle, just 3.4 centimetres wide. The frequency of the radio waves as received at Madrid rose a little whenever the antenna was moving towards the Earth and fell as it moved away.

This Doppler effect, well known to scientists and traffic cops, was extremely small in this case – a frequency shift of one part in 10 billion. The flight operations team had not expected Giotto's spin to be detectable in the weak signals so early in the operations. They were reassured that the spin rate was close to what they expected. And Pasadena's success meant that Darmstadt could rely on the Doppler effect to detect small motions in the spacecraft in response to commands from the ground. They were no longer completely blind.

'Who needs telemetry when you've got Doppler?' said Howard Nye.

Tuesday, the second day of the reactivation, was less encouraging. In the menu of manoeuvres figured out in advance of the operations, command cycle zero was an attempt to fire Giotto's hydrazine thrusters to swivel the spacecraft, assuming that no adjustments were needed to its internal systems. The more powerful transmitter feeding Giotto's radio dish was commanded on, in the hope that its beam would sweep across the Earth.

The consoles at Madrid and Darmstadt showed not even a flicker of activity. It was like waiting for the phone to ring. After several hours the command went out to switch Giotto's transmissions back to the small antenna. Pasadena then told Darmstadt that the Doppler signals looked just the same as the day before. The spacecraft had not budged.

The mission controllers reconsidered their strategy. All the Earth- searching manoeuvres they had listed meant nothing if they could not command the thrusters. They decided to keep the small antenna switched on, and use the Doppler effect to check directly if the spacecraft was responding. On the Wednesday they tried this, only to confirm that the thrusters were not working.

Equally depressing was a test of the despin motor of the radio dish. When this was commanded on, there should have been a slight but detectable change in the rate of spin. There wasn't. If the despin motor could not be reactivated, it would not keep the radio dish trained on the Earth during subsequent operations. Giotto would be for all practical purposes useless – even if it stopped ignoring orders.

By the Thursday hurriedly written software enabled Darmstadt to analyse for itself the Doppler effect in the signals, as discovered by NASA. It reproduced the American results retrospectively, although in real time the small team continued to obtain the analyses by voice and fax from Pasadena. But the technical success seemed hollow. The third attempt to manoeuvre Giotto was as ineffective as the others.

On that Black Thursday, 22 February, many in the team feared that Giotto was dying out there in the desert of space. They might have only one more chance of resuscitating it. The spacecraft engineers studied the signs and symptoms, and pored over the circuitry that relayed the commands to the various mechanisms. They identified places where faults could explain Giotto's inability to fire its thrusters, and others where the order to despin the radio dish could be blocked.

In one place the possible trouble-spots coincided. A single fault in a chain of circuits linking attitude control and the distribution of power could account for both difficulties. The best bet was to assume that this was the seat of the trouble, and use a spare circuit for rerouting the signals within the spacecraft. Commands sent the following day instructed Giotto to bypass the suspect link.

Everyone at Darmstadt was ready on Friday for an all-or-nothing effort to coax the spacecraft into action. The vital despin motor was commanded on and a subtle Doppler change seen by NASA showed that it was turning at last. While waiting for news on that response, the mission controllers sent new instructions to the thrusters, to begin altering the spacecraft's attitude.

In a brief change of fortune, everything seemed to go right. Sixteen minutes after the first thruster command went out, and only four minutes after the subsequent signals from Giotto first reached the Earth, Morley could tell from the Doppler analysis at Darmstadt that the spacecraft was slewing. After another four minutes the spacecraft had completed a relatively quick attitude change, altering its tilt in relation to the Sun. Next, Giotto was supposed to begin a swing towards the Earth.

Hearts sank again. For ten minutes Pasadena reported no hint in the Doppler analysis of any response. Then the signs of movement reappeared and continued strongly for the rest of the eighty-minute manoeuvre. The flight dynamicists understood the apparent lull at the start. During those anxious minutes, Giotto's turning motion was almost exactly at right angles to the direction of the Earth –

the worst situation for using Doppler. But it gave Morley and his team a strong clue to the spacecraft's orientation, for planning the next day's manoeuvres.

On Saturday Giotto received orders to transmit with its radio dish. The thrusters were set going in a bid to sweep the radio beam across the Earth.

'There she is again!' Nye said, as data suddenly flashed on the monitors.

Madrid was receiving Giotto loud and clear, no longer by a feeble carrier wave but by robust signals carrying data on the spacecraft's condition. They did not last long, because Giotto was still swinging. The 'halt' signal went out right away, but by the time it reached the spacecraft the beam was already pointing twelve degrees away from the Earth. The controllers inched it back in three small steps until the radio dish was locked on the Earth. The signals were poor, though, in respect of the telemetry data sent by the spacecraft. There were many obvious errors.

'Are we going to have to fly to Grigg-Skjellerup with bad telemetry all the way?' Nye wondered.

The signals that did make sense were worrying. The console screens were a mass of red lettering, signifying departures from the nominal. 'The spacecraft is very hot, especially at the top,' an operations engineer reported.

Giotto was in a sector of its orbit taking it close to the Sun, where the solar rays were more than 50 per cent more intense than at the distance of the Earth. The overheating was aggravated by damage to the power dumpers that were supposed to get rid of excess power from the solar cells. The 'top' end of the spacecraft carrying the antennas was at that time tilted towards the Sun.

Fresh commands went out to change the spacecraft's angle to the Sun by fifty degrees of arc. This reduced the power output from the solar cells and also gave the 'top' an opportunity to cool down. The scientific instruments at the 'bottom' now bore the brunt of the solar heat. They remained switched off, so perhaps they could stand the punishment.

Just when it seemed that Giotto was being brought under control, a new emergency arose. After only an hour in full flood, the stream of data from Giotto stopped altogether. The previous bad data had been a warning of imminent failure in a component that was probably overtaxed by the high temperatures. Tests showed that both transmitters were still working but Giotto's data-handling system had packed up. The team at Darmstadt was in for a trying weekend.

Had all the nursing and coaxing been in vain? Through Saturday night and Sunday morning that seemed a dire possibility. Once again the engineers had to identify the fault and work around it. They were helped this time by diagnostic tests on the spacecraft that eventually pinned the blame on a terminal unit. It was letting commands into the spacecraft but blocking the transmission of telemetry to the Earth.

The engineers and operations team planned an intricate rearrangement of Giotto's internal electronics to use the spare terminal unit. On the Sunday afternoon the Madrid station beamed the necessary instructions to the spacecraft. When the reconfiguration was complete, the data streams reappeared on the consoles at Darmstadt, unblemished by the former errors.

On Sunday evening, 25 February, six days and six hours after the first wake-up signals had gone out, the flight operations director declared Giotto reactivated. Its survival, despite the damage at Halley and the subsequent overheating, was more than just a tribute to Europe's engineers. With much-appreciated help from NASA at Madrid and Pasadena, the European Space Agency had accomplished a 'first' in space technology, by reviving a spacecraft after a long hibernation.

Until then, planners of unmanned missions into deep space thought that, even in an interplanetary flight lasting several years, you had to keep a control crew and ground stations on hand to communicate with the spacecraft, say once a week. Although not designed for this mode of operation, Giotto put up with four years of neglect. There was money to be saved by hibernation in future missions, starting with the second leg of Giotto's journey to Comet Grigg-Skjellerup.

Darmstadt settled down to a less frenzied period of tracking the homecoming spacecraft as it closed with the Earth at almost a million kilometres a day, and fidgeting with Giotto's attitude as it made its closest approach to the Sun early in March. The temperature throughout the instrument deck went above the design limit of 50 degrees Celsius, so the planned checkout of the scientific payload had to be deferred until it was cooler.

Late in April, members of the science working team gathered once more at Darmstadt with their experimental ground-support equipment. Like the parts of Giotto itself, the ground computers used for handling the spacecraft and the real-time data from its scientific instruments were elderly or defunct. The rapid progress in computing during the 1980s had left them looking, as Gerhard Schwehm said, 'like something out of the Stone Age'. Devising interfaces between the modern equipment at the European Space Operations Centre and the old Giotto systems was a tiresome task during the extended mission.

Since the buffeting by dust and electrical discharges during the encounter with Halley's Comet, the scientific instruments in the spacecraft had long-since passed their sell-by dates, and suffered wide changes in temperature, culminating in the latest insult. The results of the checks showed remarkably little change from the impressions formed right after the encounter with Halley. All the experiments were to varying degrees functional except for Dieter Krankowsky's neutral mass spectrometer, struck dead by Halley, and Uwe Keller's camera, blinded.

Special efforts went into reviving the camera, using a spare version of the instrument to simulate the responses to the commands going out to its sibling in space. Like the spacecraft, the scientists were sweltering. In a heatwave at Darmstadt, an eight-man camera team and its elderly equipment were crammed in a mobile cabin without air-conditioning. Schwehm watered the cabin's roof and walls with a garden hose to try to lower the temperature inside.

The tests confirmed that the camera had lost its baffle – the tube that pointed in the direction of its field of view and screened it from the Sun. No matter how the camera rotated, the baffle cast no shadow on the solar cells, which would have affected their power output had it been in place. The baffle was not essential for obtaining images and the internal parts of the camera were working. They pointed the camera towards the Earth, Jupiter and a bright star. Then they scanned it across the sky. There was never an image.

It was as if a photographer had left his lens cap on. The most likely explanation was that a piece of the broken baffle had lodged itself across the optical aperture. Twisting and turning the instrument did not shake it off. A proposal to throw it free by making Giotto spin faster was ruled out because of possible damage to the spacecraft's despin motor.

The lack of success with the camera seemed like a blow for the proposed mission to Comet Grigg-Skjellerup, which was still not formally approved. The decision on whether or not any further money should be spent in attempting to conduct any scientific observations at the encounter was postponed until 1991. Taking the Giotto Extended Mission step by cautious step, the European Space Agency authorized expenditure on the Earth swingby and the retargeting of Giotto towards its second comet.

'Gravity assist' or 'slingshot' were other names for the technique whereby a spacecraft flying close to a planet could exploit its gravity to speed itself up or slow down, and change its orbit. The only rocket power needed was that of the small thrusters used to steer the spacecraft through the encounter. The Earth swingby would boost Giotto's speed by more than twice as much as did the onboard rocket motor, the Mage, when it helped the spacecraft on its way to Halley. Although it was the space engineer's free lunch, swingby was not a human invention, and a comet probe so guided had an affinity with its targets.

Nature has used swingbys for billions of years to impose a little order on the Solar System, and especially on the unruly comets that infest it. While the major planets circle the Sun in neat, secure orbits, the comets rush through the traffic threatening to cause an accident. A near miss is always much more likely. While a comet is far too light to influence a planet's orbit significantly, a close encounter with a planet will alter the comet's path around the Sun, mildly or severely.

The comet cannot gain or lose speed in relation to the planet. Any acceleration from the gravitational pull of the planet on the way in is cancelled by the backward drag on the way out. What matters for the long-term orbit is the comet's speed and direction in relation to the Sun. These change in the 'swing' of the swingby, as the comet's path swerves towards the side on which the planet lies as it passes.

A comet crossing the wake of a planet can be pulled round into a fast orbit similar to the planet's own. But if it overtakes a planet and passes so close in front of it that the swingby sends it back on its track, the comet loses most of its speed in relation to the Sun. In 1886 Comet Brooks 2 made a U-turn in front of the giant planet Jupiter. It cost the comet so much of its energy of motion around the Sun that the period of its orbit was slashed from twenty-nine to seven years.

Astronomers who predict reappearances of comets like Halley or Grigg-Skjellerup have to take account of planetary influences. Though always troublesome for the predictors, the changes are usually unspectacular, being due to distant encounters. Even so, most comets experience drastic orbital changes at various times during their existence. Swingbys govern the evolution and fate of comets, and in some cases put them into compact orbits like Grigg-Skjellerup's.

The kick-off occurs when the gravity of a passing star topples a swarm of comets from their distant deep-freeze in the Oort cloud that supposedly surrounds the Solar System. Those that gain energy from the encounter are expelled from the Sun's gravitational domain without more ado. Others lose energy and drop towards the Sun. When they reach the inner Solar System they are travelling very fast, and belching gas and dust for the first time. Sometimes they hit the Sun, but usually they swing around it and head outwards again.

The 'new' comets would return to their starting places, if they did not feel the gravity of one planet or another and either gain energy or lose it. Massive Jupiter is the star player, with the most chance of affecting the flight of a comet. Even a distant swingby can speed up a 'new' comet sufficiently to expel it for ever into interstellar space. Other new arrivals slow down, in orbits that will bring them back for at least one more appearance close to the Sun. The return may be after a million years, or more quickly, depending on the loss of speed. The slower a comet travels, the sooner it comes back, because its orbit is smaller.

Each returning comet has further brushes with the planets, which cause them either to gain speed and shoot off, or to lose speed and adopt smaller, briefer orbits. By drastic swingbys the Solar System maintains a stock of about a hundred comets on compact orbits which bring them back to the Sun's vicinity at intervals of less than 200 years. Halley's Comet, on a seventy-six-year orbit, belongs to this group, but most are on much briefer orbits.

If repeated swingbys still do not evict a comet from the Solar System, its energy erodes until it settles into a comparatively safe orbit, tucked inside Jupiter's. Grigg-Skjellerup is one of a couple of dozen 'old' comets that have finished up here, with orbital periods of around five or six years. Comet Encke has the briefest, at 3.3 years. All go the same way around the Sun as the planets do.

These comets in the smallest orbits visit the Sun often and use up their stocks of ice. They remain at risk of expulsion by a close swingby, but are likely to expire before that happens. They probably evolve into dark asteroids orbiting close to the Sun. The game ends, after millions of years, with half of the ex-comets being booted from the Solar System, and half colliding with a planet. The Earth is the most effective asteroid-trap this side of Jupiter.

That Giotto's second cometary target should be one of the old, tight-orbit, short-period comets was almost inevitable. Only these offered frequent appearances with high predictability. Yet even without the operational advantages, many comet scientists would have wished to send a comet probe to Grigg-Skjellerup or one of its kin, to look for direct evidence of ageing.

As for manmade swingbys, of the kind that was to send Giotto to meet Grigg-Skjellerup, space navigators always had to take account of lunar and planetary influences on the motions of far-flung spacecraft. In the first inventive use, the American Mariner 10, dispatched in 1973, swung by the planet Venus and altered its orbit to accomplish three encounters with the planet Mercury. The Voyagers went through the outer Solar System like a football passed from one planet to the next. The five lunar swingbys that sent ICE to its encounter with Comet Giacobini-Zinner in 1985 was the most complicated succession of free lunches ever enjoyed by a spacecraft.

The bookkeeping laws of the Universe were not violated. Any energy gained by a spacecraft at a planetary swingby was subtracted imperceptibly from the planet's own motion. Thus Trevor Morley calculated that Giotto's acceleration in its swingby of the Earth, to reach Comet Grigg-Skjellerup, would slow down the planet enough to put it one millimetre behind schedule after 100 million years.

This was to be the first purposeful use of the Earth for such a manoeuvre. The standards of excellence set by the Americans put Morley's flight dynamics team on their mettle, as the time drew near.

At Giotto's reactivation on its approach to the Earth, its speed differed by only one part in 3000 from the 37.4 kilometres per second predicted for that day from the circumstances at switch-off. After four years and 3 billion kilometres of travel, the spacecraft was only 14,000 kilometres from where it was expected to be. Such minute discrepancies were not insignificant for cosmic marksmanship of the quality needed at the Earth swingby.

The knack was to achieve just the right acceleration into a new course, heading in just the orbit needed to intercept Comet Grigg-Skjellerup. Going too close to Earth, the spacecraft would change course too much; staying too far out, it would swerve too little. The reckonings worked backwards in time from the encounter with the comet, far out in space and two years later. They also worked forwards from Giotto's rapidly changing position and speed as it approached the Earth.

To match the start and the finish, the swingby had to change Giotto's orbit around the Sun from 10 months to 13.5 months, by a swerve of sixty-four degrees and a resulting boost of 3.1 kilometres per second in its speed in relation to the Sun. But to aim around the corner correctly, the controllers at Darmstadt needed to pass Giotto through the eye of an invisible needle 22,730 kilometres above the Earth's surface, at a point forty degrees south of the Equator and at a speed of 6.3 kilometres per second relative to the Earth.

Small corrections to the orbit began two weeks after reactivation, but the most precise control was left till later. The flight dynamics team was having to fidget every few days with Giotto's attitude, first to keep it cool and then, as it edged away from the Sun, to keep it warm. Each time the thrusters fired to slew the spacecraft they confused the aim by slightly altering the trajectory. Altogether sixteen attitude adjustments were ordered between reactivation in February and the start of near-Earth manoeuvres in June.

Giotto's own navigational aids helped in checking its orientation. The Sun sensor was working, and although the damaged star mapper was dazzled by the Sun it could see the Earth plainly enough on the shady side of the spacecraft. Radio communications became easier as the distance diminished. After a large manoeuvre on 16 June, which put the spacecraft into its preferred attitude for the swingby, the small look-anywhere antenna replaced the radio dish. The star mapper lost sight of the Earth. Trevor Morley waited anxiously for more than a week until the Moon came into view in the star mapper, and he could be sure that the spacecraft's attitude for the swingby was as it should be.

A 34-metre dish at NASA's ground station at Madrid checked Giotto's track and speed four times, using the feebler signals of the small antenna. Computers at Darmstadt refined the aim point for the swingby to match the more precise knowledge of Giotto's incoming path. Responsibility for tracking Giotto through the swingby passed to the German space agency's 30-metre dish at Weilheim in Bavaria, and the European Space Agency's 15-metre dish at Perth in Australia.

The final aiming, by the last small speed change, was set for Saturday 30 June, forty-eight hours ahead of the swingby event. Giotto would still be 600,000 kilometres away, beyond the orbit of the Moon. Thereafter, nothing would halt its headlong rush past the Earth, and by the following Tuesday it would be scampering off into the depths of the Solar System.

There would be no second chance. Any human error or failure of the machinery would consign Giotto to the limbo where old interplanetary probes wander aimlessly for ever through empty space. The tension made everyone subdued at mission control, but Morley was confident and so, he sensed, were the operations engineers.

At 10.00 zulu, or midday by German time, the crucial command went out to Giotto. It told a thruster in the side of the spacecraft to fire once in every rotation. The thruster's impulses, all pushing the same way, nudged the spacecraft to improve the aim. The calculations called for seventy-four minutes of this 'radial-pulsed' thruster action. Radio observations confirmed that the result of the last orbit correction was excellent. The thruster had changed Giotto's speed by 29.2 metres per second instead of 30 as intended, and had acted just half a degree away from the intended direction. The discrepancies were trivial for an encounter proceeding at more than six kilometres per second.

Susan McKenna-Lawlor and Fritz Neubauer turned the Earth swingby into a scientific mission. Any attempt to receive data in real time from Giotto's instruments was ruled out on technical grounds, but these two had recorders. The project manager agreed that the instruments could be switched on to observe energetic particles and magnetic fields in the Earth's vicinity. The controllers would order Giotto to transmit the recorded data during a lull in operations after the swingby.

Apart from its sightings of the Earth and the Moon in the star mapper, the first that Giotto knew of its homecoming was the curved space of the Earth's gravity, telling it to start to swerve. The perturbation of its orbit first became perceptible about three days before the swingby. It would have happened even if Giotto had been an inert meteorite, but in fact it was the very first scientific spacecraft to visit the Earth from interplanetary space.

Its veering track took Giotto on an unusual route through the magnetosphere, the invisible zone with a comet-like tail where the solar wind bottled up the Earth's magnetism and denied it the same long reach as the planet's gravity. Those with long memories recalled that Giotto's genesis included a proposal for a dual mission to the Earth's magnetosphere and then to Halley. It was happening after all, though in the reverse order. When Fritz Neubauer's magnetometers began operating, more than ten hours before the closest approach to the Earth, they recorded the mildly fluctuating magnetism of the solar wind on a quiet day.

On Monday 2 July 1990, Giotto was streaking in from the Earth's sunward side, five years to the day since it first rode into space from Kourou in French Guiana. At a little before 04.00 zulu, Susan McKenna-Lawlor's instrument EPONA began to sense an increase in the counts of energetic particles. At 04.49, the magnetometers felt a sharp intensification in the solar wind's magnetic field, as Giotto passed through the Earth's bow shock at a distance of 85,000 kilometres from the planet. This was where the planet began to hamper the solar wind's smooth flow.

For nearly two hours the magnetism remained nearly constant in strength, in the zone called the magnetosheath, but its direction shifted as the magnetism of the solar wind draped itself around the obstacle presented by the Earth. At 06.44, the magnetometers recorded brief fluctuations as Giotto entered the Earth's own magnetic field at the magnetopause, 62,000 kilometres out from the planet. The magnetism quickly grew stronger as Giotto closed with the Earth. Its course carried it through part of the Van Allen radiation belt, where charged particles remain trapped for years on end in the magnetic field, making a thick girdle around the Earth. EPONA's counts of particles in the belt peaked at 07.30.

All the while, the spacecraft was accelerating in the Earth's gravity, to cross the planet's wake and pass to the night side, away from the Sun. For Giotto it was plain sailing along a curved track in the shape of a hyperbola. It came closest to Earth over a point forty degrees south of the Equator. This was convenient for the European Space Agency's ground station at Perth in Western Australia, but tracking the spacecraft required a careful reckoning that took account of the Earth's rotation. As seen from the ground, Giotto made a prolonged pirouette over Australia, as its high speed eastwards overcame for just a few hours the normal effect of the turning Earth, which drove all cosmic objects westwards across the sky.

It was a dark winter's evening in Australia and the European Space Agency had alerted astronomers to look out for Giotto, glinting in rays of the Sun beyond the Earth's shadow. No one saw it or managed to photograph it as a streak of light. At 10.01 plus 18.15 seconds zulu the swingby climaxed in the closest approach. Giotto passed through its needle's eye 22,731 kilometres above the pre-ordained spot in the ocean, 700 kilometres south of the Western Australian coast.

Giotto was swerving most sharply in the Earth's gravity, with its track curving to the left and upwards as viewed from the north. From then on it gradually lost the extra speed in relation to the Earth that it gathered on the way in, but it kept its gain in relation to the Sun that the swingby was meant to accomplish. At 14.30 Giotto entered the comet-like tail of the Earth's magnetosphere, in a transition signalled by a switch in the direction of the magnetic field. By then the spacecraft was directly over the Philippine Sea and 80,000 kilometres away on its outward track.

It passed out of the Earth's magnetic domain, in the northern part of the tail of the magnetopause, at 21.20 and 180,000 kilometres off. The magnetic instruments then recorded very strong turbulence in a region of the magnetosheath never previously visited by a spacecraft. Corresponding observations by EPONA showed marked variations in the fluxes of energetic particles.

Not until 13.49 on the following day, 3 July, did Giotto cross the Earth's bow shock again, outward bound. The spacecraft emerged into a sector of the solar wind magnetized in the direction opposite to that prevailing when the spacecraft entered the region of Sun-Earth interaction thirty-one hours earlier. Variable fluxes of particles were still being encountered until the observations ceased at about 20.00 on 5 July.

Giotto's 'mission to planet Earth' was a success. Although only a small part of its payload was operational during the Earth swingby, the variety of phenomena that Neubauer and McKenna-Lawlor recovered from their data-recorders were important psychologically as well as technically. They proved that Giotto was not just a damaged machine, navigable with difficulty. It was a functional scientific spacecraft that might well make useful observations when it reached Comet Grigg-Skjellerup two years later, if anyone would pay to have it switched on.

Had Giotto carried sensors of the political climate, it would have registered amazing changes in the world since it left for Halley's Comet. Iraq was menacing its neighbour Kuwait, so peace remained elusive, but the Berlin Wall was down. Joyful Europeans began healing the scars of forty-five years of ideological and military partition, while the brooding Soviet Union looked more like a lame beggar than a nuclear warlord.

CHAPTER TEN

GIOTTO'S SECOND MISSION

GIOTTO scurried away from the Earth on its new gravity-boosted course towards Comet Grigg-Skjellerup, and mission control had three more weeks of work. Trevor Morley's team wrote the commands for turning Giotto through 110 degrees and settling it in a thermally comfortable attitude for its next long sleep. Its radio dish would remain pointing at the Earth for the time being. The commands went out on 7 July 1990, five days after the Earth swingby.

By 16 July, the trackers could see that Giotto was already on a good course for meeting Grigg-Skjellerup, but the mission planners wanted to aim as accurately as possible at the expected position of the comet's nucleus. The odds against actually hitting an object perhaps a kilometre wide were very long, but this aim-point gave the best chance of a close approach. Only a very small course correction was called for, using a thruster to alter the spacecraft's speed by half a metre per second.

After another seven days of tracking, Darmstadt put the spacecraft into its sleeping-beauty mode for the second time in its extended career. The last instruction to switched off the transmitter, while leaving the receiver on, against the day in 1992 when another of NASA's radio shouts might awaken Giotto again.

The mission, too, went back into hibernation, as far as the European Space Agency's accountants were concerned. Manfred Grensemann, project manager of the Giotto Extended Mission, aided by Gerhard Schwehm as project scientist, had yet to persuade the decision-makers that the operations visualized for 1992 were worth funding. Was Giotto fit for serious science at its next comet? For the agency's advisers, and the national delegates who would have to give the final authorization, the answer depended on the state of the instruments.

The loss of the camera, long regarded as the touchstone, might easily have put paid to the mission. But the outcome of a formal scientific evaluation, carried out a month before the Earth swingby, was surprisingly heartening. Experts who had expected to advise against the mission changed their minds and became enthusiastic about it.

The payload check after reactivation showed five experiments to be essentially undamaged: Fritz Neubauer's magnetometers, Susan McKenna-Lawlor's energetic-particles analyser, Anny-Chantal Levasseur-Regourd's optical probe, part of Alan Johnstone's plasma analyser, and Jochen Kissel's dust mass spectrometer. Tony McDonnell's dust-impact detectors were slightly damaged. Part of the Johnstone plasma analyser designed for detecting particles of the solar wind was defective, but the other part, better adapted to particles from the comet, was functional.

In Henri Rème's plasma analyser, Axel Korth's chemical instrument was defunct, and the main electron detector was not perfect. A puncture due to by Halley dust, impaired its calibration and created the risk of a short-circuit which troubled the engineers. Hans Balsiger's ion mass spectrometer had lost the unit called HERS designed for looking at the outer regions of a comet, while HIS for the inner regions was in good shape. Altogether eight out of the original ten experiments were operable in whole or in part.

For Balsiger, and Kissel too, their instruments' survival meant little. They could not function properly at Comet Grigg-Skjellerup. Tailored for Halley, they were designed to scoop up material coming from a pre-planned direction at a very high speed. Giotto's orientation and relative speed at Grigg-Skjellerup would be completely different, so the ion mass spectrometer and dust mass spectrometer could not work as intended. Kissel did not even go to Darmstadt for the checkout. He provided command sequences to verify that the instrument was functioning, but declared that he did not want to take part in any Grigg-Skjellerup mission.

As a principal investigator with little direct interest in its possible extension, Balsiger was appointed to the jury advising on the future of the project. So was Dieter Krankowsky, whose neutral mass spectrometer was defunct. The other members represented the agency's space science advisory committee and Solar System working group.

When he met them at Darmstadt on 29 May, Grensemann reported on the condition of the spacecraft as a whole. He told how the defects encountered at reactivation were cured, though with the loss of back-up systems. The solar cells and the transmitters were in good condition, while other parts including the star mapper and several pieces of Giotto's thermal insulation system were damaged.

With the batteries assumed to be dead, temperatures and power supplies would be close to operable limits at the encounter with Grigg-Skjellerup.

Schwehm presented the results of the payload checkout. He had to convince the sceptical scientists that the surviving instruments made the second mission worth while. In the end, the experts were persuaded. They concluded that the dust measurements would be valuable both for studying the evolution of comets, by comparison with Halley, and for reconnoitring the conditions to be faced by future comet missions to short-period comets akin to Grigg-Skjellerup, while the surviving magnetic and particle instruments were capable of making a 'major contribution' to the study of interactions between the solar wind and comets.

Grensemann and Schwehm were convinced that the full mission would eventually be funded, and their part-time work to carry it forward never ceased. After the expert review they could point out that Giotto would make valuable scientific observations at its second comet, at a cost of less than a tenth of the original Halley mission, including the money already spent on the reactivation and swingby.

'We have a fascinating mission,' Grensemann insisted. 'We have the first chance to use a single spacecraft dedicated to comet exploration to make a comparison between two comets – an old one, Grigg-Skjellerup, and a relatively young one, Halley.'

While Giotto had been travelling ever since their encounter, so had Halley's Comet. Five years into its journey back into the depths of space after its visit to the Sun, sensitive telescopes were still tracking its faint nucleus at a distance of more than 2 billion kilometres. The comet had lost its bright head and tail because the Sun's rays were too feeble to warm its ices sufficiently to generate fountains of gas and dust. Its surface temperature was minus 200 degrees Celsius.

Astronomers at the European Southern Observatory in Chile were therefore astonished to see an explosion on Halley. On 12 February 1991 the comet appeared 300 times brighter than expected, and a cloud of dust 200,000 kilometres wide surrounded the previously bare nucleus. A month later the cloud was wider still and changing in appearance from night to night. Violent outbursts had often been observed in comets before, but never at so great a distance from the Sun.

This was no small puff. David Hughes of Sheffield calculated that the outburst was equivalent to about a quarter of what the comet emitted during an entire visit to the Sun. He suggested that Halley had collided with a meteorite, and went on to argue that such impacts were a neglected feature of the evolution of comets.

The event that caused the outburst was certainly short-lived, but other experts preferred to imagine some kind of internal quake or explosion. One of the explanations on offer was that heat absorbed during the recent visit to the Sun slowly reached the comet's interior, and there triggered a sudden change in the arrangement of the water molecules in very cold ice. If its temperature rose above a critical level, the ice itself would release more heat and so warm adjacent ice, in a chain reaction. Gas would also escape from the warmed-up ice. By blasting dust into space, it could have created the observable cloud.

Whatever the explanation, Halley's eruption showed that scrutiny by high-powered telescopes and visiting spacecraft had not robbed the famous comet of its capacity to surprise earthlings. The event gave comet madness a new lease of life. Advertisements by the Scientific Forecasts Society in London newspapers claimed that the explosion had dislodged Halley from its orbit and that it was coming back towards the Earth – 'Unexpectedly! Now!'

People were older too, a dozen years after the conception of the Giotto mission. This was a factor taken into account in preferring Grigg-Skjellerup to comets that the spacecraft could have reached only after more protracted interplanetary flights. And no one spent the years of Giotto's hibernation waiting for its uncertain revival. There were other duties and opportunities.

Several of the spacecraft design and engineering team at British Aerospace in Bristol, who had worked on Giotto, were engaged in a major new project for the European Space Agency: the Polar Platform. Due for launch in the late 1990s, this was to be a large unmanned spacecraft for Earth observation, conceived in connection with the US-led International Space Station.

Many of the Giotto scientists were occupied with other missions, including the Ulysses solar-polar spacecraft launched in 1990, and the five spacecraft of Europe's Soho/Cluster cornerstone project due for launch in 1995. Soho was to be stationed in space at a distance of 1.5 million kilometres on the sunward side of the Earth, to observe the Sun itself and the solar wind, while Cluster would be a set of four identical spacecraft orbiting the Earth to make simultaneous observations of the interactions between the solar wind and the Earth's magnetism.

In Ireland, Susan McKenna-Lawlor was a co-investigator for experiments in both Soho and Cluster. She had already sent instruments to Mars. Soon after Giotto's encounter with Halley, her colleagues at Katlenburg-Lindau had alerted her to the possibility of flying a particle detector on the two Soviet Phobos spacecraft bound for Mars and its moons and she succeeded in placing an Irish instrument aboard them. Although contact with the spacecraft was lost in the vicinity of the moon called Phobos, she obtained pioneering results on the particle radiation close to Mars itself.

McKenna-Lawlor was then invited to take part in planning a follow-up experiment for the Soviet Mars-94 mission. She was to contribute a successor to the Phobos instrument, with more mass and power allotted to it. She became a co-investigator for three other experiments, and for two of these she was named 'official manufacturer'. The break-up of the Soviet Union at the end of 1991 left Mars-94 with severe financial problems, but the preparation of its experiments continued. McKenna-Lawlor was also co-investigator for an energetic-particles experiment in the proposed Russian Universal Space Platform, and for a French radio experiment in a NASA satellite called Wind.

She was still professor of physics at St Patrick's College, and her emblem was still Epona, but the Celtic goddess was now part of the logo for McKenna-Lawlor's company Space Technology (Ireland) Ltd in Maynooth's industrial park. Promoted by a Dublin financier, the company offered flight qualified hardware, onboard data processing, ground-support equipment, environmental testing and numerical modelling. It could claim that it was building space instrumentation for missions of the European Space Agency, the US and Russia. Spinoffs included work on racing cars and shipboard receivers for satellite communications.

The growth in space science and technology at Maynooth meant that young Irish men and women no longer had to look only abroad, as McKenna-Lawlor herself first did, for opportunities in space. But it also meant that even for one as passionate as she, about her experiment in Giotto, the extended mission was strictly a sideline.

Nobody had the heart to kill poor Giotto. The survival of the camera had been a criterion of whether the second reactivation for the encounter with Comet Grigg-Skjellerup would be worthwhile. When that instrument remained stubbornly inoperable, the scientists successfully argued that a quorum of other instruments existed. They found a useful basis for comparison. While Giotto at Grigg-Skjellerup might be a pale shadow of Giotto at Halley, it should be much more effective than the US-led International Cometary Explorer mission to Comet Giacobini-Zinner in 1985.

At the headquarters of the European Space Agency in Paris there were moves to make further operations with Giotto an optional extra to the science programme, to which interested member states could subscribe if they wished to do so. That would relieve the pressure of the unscheduled mission on the science budget. But in 1991 new accounting principles reduced the overheads of the science programme. NASA continued to make the facilities of its Deep Space Network available on a basis of quid pro quo. As a result the science policy committee was able to decide in June 1991 that the Giotto Extended Mission should go ahead, as part of the agency's regular, mandatory programme.

The target comet came sailing in to pay its respects to the Sun, on an orbit that would converge nicely with Giotto's. Astronomers checked Grigg-Skjellerup's movements at its visit in 1987 and continued until 1989 when it was far away and hard to see. Their diligence helped in aiming the spacecraft at the Earth swingby and also ensured the prompt detection of the comet at its next return. A German-Spanish observatory at Calar Alto in Spain picked it up on 9 September 1991.

The comet's name commemorated two amateur astronomers. John Grigg, a teacher of singing in New Zealand, found it in 1902; Frank Skjellerup, an Australian telegraphist working in South Africa, rediscovered it twenty years later. After Grigg spotted the round smudge in the winter sky near Auckland, in July 1902, a waxing Moon prevented him observing its motions for more than ten days. Predictions of the orbit were therefore vague. When Skjellerup saw the comet in May 1922, retrospective computations proved it was the same as Grigg's.

Later a Slovakian astronomer, Lubor Kresák, linked Comet Grigg-Skjellerup with a sighting in 1808 by France's most famous comet hunter, Jean-Louis Pons. Encounters with Jupiter in 1809 and 1845 altered the orbit drastically. More distant brushes in the 20th Century kept varying by a few weeks the period between the comet's visits to the Sun.

A meteor stream called the Sigma Puppids was supposedly a trail of Grigg-Skjellerup's dust. Scientists operating a large radio telescope in Puerto Rico, who had managed to detect the nucleus of the comet by radar, said it was at least 400 metres in diameter. David Hughes of Sheffield calculated that the figure should be 460 metres, if Grigg-Skjellerup were like Halley but just smaller. But if it were a dying comet, dribbling gas and dust from active areas scarcer even than Halley's, the nucleus could be more than a kilometre wide. It was unlikely to rival Halley's mean diameter of eleven kilometres.

Whenever Grigg-Skjellerup came closest to the Sun, at roughly the distance of the Earth's orbit, it brightened and faded quickly like a burglar cautiously flashing a light. The faintest stars visible to the naked eye were 10,000 times brighter than the comet. Even electronic images from large modern telescopes showed only its round head, or coma. Grigg-Skjellerup was a comet with no visible tail.

The stage directions for Giotto's second encounter were quite different from the spacecraft's nearly head-on rush into Halley. At Grigg-Skjellerup they read, 'Enter, pursued by a comet.' Seen from the Sun's viewpoint, the Earth went by first, from right to left, passing very near to the scene of the impending encounter but three months ahead of the event. Giotto followed it on a slightly wider but converging orbit, and at a speed similar to the Earth's. Grigg-Skjellerup appeared a little later on the same side, but lower down and farther from the Sun.

The comet was accelerating on a rising track that slanted at 21 degrees to Giotto's orbit. It would overtake the spacecraft in centre-stage and engulf it briefly. At that moment the comet would be travelling at thirty-eight kilometres per second, and Giotto at thirty-one. Continuing onwards and upwards, Grigg-Skjellerup would sidle most closely to the Sun twelve days after the encounter, and exit top left.

From Giotto's point of view the comet would rush up at it from the south, at fourteen kilometres per second. Although brisk, the relative speed was much less than the sixty-eight kilometres per second at Halley. That was just as well. Giotto's dive into Comet Grigg-Skjellerup would win no points for gracefulness.

Custom-built for the Halley encounter, Giotto fitted into the Solar System's geometry of 13-14 March 1986 like a key in a lock. The beam angle of the radio dish, squinting at 44.2 degrees to the spin axis in to suit the Earth's whereabouts seventy-six months before, was inappropriate for Grigg-Skjellerup. The radio dish would still have to point to the Earth, and the solar panels wrapped around the spacecraft's drum had to stay square on to the Sun's rays, lapping up as much power as possible.

The only way to meet both requirements at the same time was for Giotto to enter the comet with its spin axis tilted at sixty-nine degrees to its direction of motion relative to the comet. In other words, instead of going in bumper shield first the spacecraft would be turned almost sideways on.

'At the next encounter,' Trevor Morley warned, 'the solar cells will bear the brunt of the dust impacts.'

The belly-flop would expose the instruments and transmitters to danger too. It would have been fatal at Halley. But Grigg-Skjellerup was much less dusty and the lower relative speed meant that, grain for grain, the impact energy would be only 4 per cent of that experienced at Halley. The engineers and experimenters hoped that Giotto might survive long enough to make some useful observations.

Alan Johnstone

Manfred Grensemann

Gerhard Schwehm

Hans Balsiger

Front: Roger Bonnet (left) and Jochen Kissel
Back: Fritz Neubauer and Anny-Chantal Levasseur-Regourd

Even if Giotto refused to wake up from its second hibernation, imposed on it after the Earth swingby in 1990, nothing short of an unlikely collision with a meteorite could stop it passing near Comet Grigg-Skjellerup. Its progress through the Solar System, like the motion of the target comet itself, was as relentless as a cannon ball's. The marksmen of Darmstadt would have scant satisfaction, though, from a computed encounter unverified by observation, if Giotto did not wake up for it.

Plans for the second reactivation relied more on faith and diligence than on a cold assessment of the prospects. A contest was impending between the good luck that had attended the spacecraft at every stage of its existence, and the spirit of mischief that always lurked, ready to alarm or perplex the Giotticians. Operating at the margin of its own capabilities and that of the ground stations, Giotto had little capacity left for resisting mischief. Some systems no longer had back-ups, so further faults might be incurable. No one would be surprised or reproachful if the Giotto Extended Mission ended in silence.

In some respects, the circumstances were even less favourable than for the tense and tricky reactivation that preceded the Earth swingby in 1990. On that occasion, Giotto was 102 million kilometres away and closing, so that the range for radio communications diminished every day. This time, the spacecraft would be more than twice as far, and the distance would change very little because Giotto was dawdling behind the Earth in its new, enlarged orbit around the Sun. The round-trip time for sending a command and seeing a response was more than doubled, and signals to and from the spacecraft would arrive with only 20 per cent of their strength at the former reactivation.

On the other hand, Giotto's shifting position in the sky, and especially its attitude in orbit, were known more accurately than in 1990. Fully a quarter of the generous stock of hydrazine fuel with which Giotto first left the Earth seven years earlier remained available for the thrusters, so manouevring should present no problems. Experience bolstered confidence too. This would not be the first time that a spacecraft had been reactivated from hibernation, and the team would be largely the same.

Howard Nye was the flight operations director. He replaced David Wilkins, who had led the flight control team since Giotto's launch in 1985 but left operations to head a new software department at ESOC. When Nye became conductor of the orchestra, with overall charge of the mission's ground facilities and flight dynamics as well as the control of the spacecraft, Antonello Morani took his place as spacecraft operations manager.

This young Italian engineer would fly the spacecraft with the Italian name through what was supposed to be its last adventure. Morani had arrived in Darmstadt five years earlier, equipped with the *bocca romana* jaw of his native Rome and a degree in electronic engineering. He joined ESOC's computer department, but a chance to learn the operations engineer's trade came when Europe's Meteosat series of weather satellites needed additional manpower. Between 1988 and 1991 he took part in three Meteosat launches.

Morani's association with Giotto began in 1990. Then, the dirty job of resuscitating a damaged spacecraft from hibernation and guiding it through an Earth swingby, with a high risk of failure, was a task that some of the more experienced people at ESOC preferred to avoid. It was a chance for a junior engineer to shine. In Nye's spacecraft operations team, Morani helped to plan and execute the manoeuvres of Giotto's first reactivation.

He compiled a manual for the 1992 reactivation, listing all foreseeable events and commands. Its thickness was a measure of the precarious state of the mission. A few pages at the beginning set out what should happen if Giotto did everything required of it at the first time of asking. The rest of the manual detailed, with intricate logic, all the variations and tricks to be tried if particular commands or manoeuvres were unsuccessful. These began with the possibility that Giotto would fail to respond to its wake-up call. Morani envisaged all manner of possible defects and mishaps, right down to the possibility that the hydrazine fuel for the thrusters had frozen in its tanks. A consolation was that the defects identified on the last occasion reduced the number of possible permutations of spacecraft 'configurations' that he had to consider.

Giotto's second mission coincided with the European Space Agency's long-waited Eureca-1 mission. A NASA shuttle flight was due to launch, that summer, a general-purpose, retrievable carrier of all sorts of scientific and technological experiments. As the more complicated mission, Eureca-1 bumped the Giotto team out of the main control room at Darmstadt, and into a support room. It was long and narrow like a submarine, barely large enough for all the people, with their consoles and chairs, who had to fit in when the time came to reawaken Giotto, on 4 May 1992.

Outside, the evening sun shone on a Germany somewhat dislocated by a rare strike in the public services. Only a thousand kilometres away, in Sarajevo, Europeans were massacring one another as Yugoslavia followed the Soviet Union into terminal dismemberment. The mission's partners in NASA's Deep Space Network in Pasadena lived on the rim of Los Angeles, which had erupted in deadly race riots just a few days earlier. By comparison, comet raiding was a gentle pursuit.

The project manager Manfred Grensemann sat near Howard Nye, together with the project scientist Gerhard Schwehm. Behind them were engineers from British Aerospace, Terry Kilvington and Bill Johnson. At the other end of the room was Antonello Morani, in brown jeans and a drab shirt, surrounded by the operations engineers of both his own and the alternate shift. Trevor Morley and his flight dynamics group were in a room next door. There were murmurs of conversation as everyone pretended to be calm.

On the screen of a small computer, adapted for the communications with Pasadena, was a list of pre-arranged commands. These changed colour, item by item, as they went to the spacecraft. Morani exchanged only a few remarks with the Giotto track controller in Pasadena, about precise timings. For the best part of an hour, NASA's Madrid station fired Darmstadt's commands at Giotto's computed position in the sky. Powered to ninety-five kilowatts and beamed by the 70-metre dish, the signals reached across 219 million kilometres of space. They jiggled electrons in the spacecraft's small look-anywhere antenna, with messages to coax some of its systems back into operation.

At 15.50 zulu, which meant 5.50 p.m. in Darmstadt, the command went out for Giotto to transmit a radio wave. The round-trip time was twenty-four minutes, and as the clocks advanced to 16.14 a hush descended on the crowded control room. There was no mock nonchalance now. Team members glanced at one another with raised eyebrows. Was their elderly and sorely-tried spacecraft still alive after another two years of solitude?

'We have receiver lock at an AGC level of minus 171 dBm.' The quizzical looks gave way to smiles as the voice of the track controller in Pasadena relayed the news from the Madrid station.

Giotto's signal was a little stronger than expected, yet 'minus 171 dBm' meant only 8 billion-billionths of a milliwatt of radio power. A fly walking on the Madrid receiver pounded it with more energy in a second than Giotto's distant whisper could supply in a million years. Well done Madrid, well done Giotto. Down came ESOC's comic posters that showed the spacecraft lying on a bed, with the inscription, 'Hibernation: do not disturb!'

More good news followed quickly. Despite the weakness of the signal, the Madrid station was already registering slight shifts of frequency, as the damaged spacecraft wobbled on its axis. The Doppler wizardry, as perfected by the Deep Space Network in 1990, would once again give mission control wonderful insight into the spacecraft's behaviour, even before full communications were established.

The flight dynamicists next door showed their sang froid on a chessboard, playing over a game in which Nigel Short had just beaten Anatoly Karpov. After the weeks of preparation for the operation, Morley's prediction of Giotto's orbit was already judged to be 'spot on'. The group had nothing to do but wait for a fax from Pasadena. It would show Doppler data in Giotto's signals at its second re-awakening.

'We're not getting any response from your fax machine,' a slightly aggrieved American voice declared on the transatlantic voice channel.

The Short-Karpov game continued until the fax came, showing the fluctuations for the first half-hour. A young Belgian flight dynamicist, Johan Schoenmaekers, required just two minutes with a photocopy of the fax and a pocket calculator, to report that the spacecraft's rate of spin was 14.928 revolutions per minute. Since Giotto was last heard from in 1990, its spin rate had increased by 0.2 per cent. Morley passed the word by the intercom to mission control.

'I'm surprised how accurate your figures are,' Morani remarked.

'Yes, how can you get so many decimal places from such a ragged-looking trace?' a bystander asked, eyeing the Pasadena fax.

Schoenmaekers shrugged. 'It's like tuning a guitar,' he said. He meant that the period of the beat from a mixing of the highest and lowest frequencies, then about eight minutes, was extremely sensitive to the spacecraft's rate of spin. Two hours later, when the flight control team tested the despin motor of Giotto's radio dish, they relied on changes in the Doppler effect to tell them it was working. The reaction to starting the motor increased the spacecraft's spin rate by less than 1 per cent, but it changed the beat period from eight to eighteen minutes.

By the morning after the resumption of operations, most of Morani's thick manual was already superfluous. Many of the dire contingencies it visualized were safely bypassed. A loss of the spacecraft's signal in the small hours was traced to a mistake at a ground station. Giotto itself seemed perky. The first commands to adjust the spacecraft's spin rate and attitude evoked a further change in the Doppler features of the signal. So the thrusters were working too. Manoeuvres to slew the spacecraft into its correct orientations relative first to the Sun and then to the Earth proceeded without a hitch, until the ground stations were locked on to transmissions from Giotto's radio dish.

Dismay came with the first telemetry. Signals from the spacecraft, which should have told the operations engineers about its temperatures, power supplies and other internal conditions, were so badly garbled they made no sense. The team was crestfallen, because in 1990 poor telemetry heralded the failure of a terminal unit inside the spacecraft. Switching to a second unit solved the problem then, but this time there was no other back-up – no more redundancy in Giotto's circuits. Even if the unit did not pack up completely, a spacecraft sending unintelligible signals could not carry out its scientific tasks at Comet Grigg-Skjellerup.

Terry Kilvington of British Aerospace looked over the shoulders of the project manager and flight operations director, and refused to believe that his beloved spacecraft was at fault. His long experience as a space engineer included an association with Giotto going right back to 1979, when his company was wondering what to do with those spare parts for Geos. And like a jolly uncle among the young controllers, he was old enough to remember the days when digital computers were troublesome brutes, quite capable of sending out million-dollar gas bills, crashing an aeroplane or garbling a message from outer space.

As he eyed the nonsense in the telemetry channels, Kilvington sensed a computer error of a very primitive kind. Something silly was happening to Giotto's signals between their arrival at a NASA ground station and their appearance, imperfectly decoded, on the display units at Darmstadt. He suggested doctoring the signals, to delay them all by the duration of one bit. A trivial alteration to the introductory lines of the data-handling program introduced a false bit at the start of the data stream, and it worked like a charm. Now the telemetry made sense.

Red-lettered lines on the display screens showed that the spacecraft was very cold, but tongue-tied it was not. Tracing the fault in the ground circuits took more than a week, until ESOC's engineers realised that a computer clock in a box at Darmstadt, which passed the incoming signals from the Deep Space Network into the mission's data-processing system, was not properly synchronized with NASA's clock.

On 8 May mission control declared the spacecraft reactivated. In the two months that remained before Grigg-Skjellerup, Giotto needed nursing like a hypothermic sailor hauled from the sea, using the onboard heaters powered by the solar cells. The scientific instruments had to be checked out again and rehearsed. And a conundrum about power supplies that had haunted the mission since 1986 now demanded a definitive answer.

Even in its heyday at Halley, Giotto had little power to spare. It needed all of the 190 watts that the solar cells then supplied. The spacecraft's batteries provided a reserve against sudden discharges due to equipment failures or comet sparks, and possible damage to the solar cells. To put the figure in perspective: Giotto ran ten experiments, looked after its housekeeping and communicated to the Earth across interplanetary space, all with no more power than one would use in a couple of electric bulbs, to light a room of moderate size.

The greater distance from the Sun at Grigg-Skjellerup would reduce the output from the solar cells by 21 per cent, or forty watts. Without the batteries, a further allowance had to be made for a safety margin against power surges – the evil consequences of which could include a shut-down of the transmissions to the Earth. There were ways of saving a few watts here and there, notably by cutting out the heaters, but the upshot was that Giotto might manage quite well at the encounter as long as no one wanted to run any scientific experiments.

More precisely, the power to spare for all the instruments was about two watts, compared with fifty-one watts for the full payload at Halley. Even if the camera and the mass spectrometers, watt-guzzlers all, were left inoperative at Grigg-Skjellerup, together with some units of the remaining experiments, it seemed impossible to manage with less than about ten watts. The mission could become scientifically pointless for want of enough power to run a flashlight.

Ruthless solutions to the conundrum were in contemplation, all prejudicial to the desires of one group of scientists or another, when Giotto came to the rescue. Despite the general opinion that its batteries were dead, based on experience with similar silver-cadmium batteries in the Geos spacecraft, the British Aerospace engineers felt confident that there would be some life left in them. After an earnest exchange of views with the battery manufacturers, seeking reassurance about possible risks including explosions, the project manager approved a check on the batteries' condition.

They worked! Despite their age, and the severe overheating they had suffered, three out of the four batteries were usable. The fact that they had been uncharged during the hibernations had evidently helped them to survive.

The batteries boosted the chances of success, and the engineers assigned them to the same purposes as at Halley. As they could take care of the power surges, they immediately made another seven watts available for the experiments. Their other role would be to keep the spacecraft running for a few minutes longer if Comet Grigg-Skjellerup destroyed its solar cells.

Relaxing the power constraints encouraged more bullish views in the assessment of experiments. Henri Rème's electron detector had a big question-mark against it because damage by Halley dust had left it punctured and liable to short-circuit. Rème was keen to make the best of it, but others feared the repercussions of possible arcing in the damaged sensor. With batteries to cushion the surges, everyone was happier about switching the experiment on.

Another beneficiary was Hans Balsiger. He was the first to admit that his ion mass spectrometer was wrongly aligned for an encounter occurring at the wrong speed. He could not promise sensible results, but he would be glad to run the experiment if the spacecraft could power it. It was added to the list.

The power demands for the experiments rose to sixteen watts, creating a new shortfall of seven watts. The mission controllers said they could save five of them by forbearing to communicate with Giotto via the uplink command channel during the encounter. At the end of all the sums, the spacecraft would go into the encounter with a theoretical deficit of two watts in its power budget, slightly reducing the safety margin. The good news was that, out of Giotto's ten original onboard experiments, seven would be at least partly active at the second comet.

As if to counteract the good fortune with the batteries, mischief struck from an unexpected quarter. Less than two weeks before the climax of the mission, an earthquake in southern California disabled the Deep Space Network's big dish at Goldstone. Mission control was counting on it to maintain communication with Giotto when it sank with the comet to the western horizon of Weilheim and Madrid, five hours after the closest approach. NASA plugged the gap with two smaller dishes at Goldstone.

Three days before the encounter, Uwe Keller and his camera team had a last go at reviving their blinded instrument. During the previous night, the controllers found that they could not prepare the camera without switching from one communications band to another. Vital work of tracking Giotto in its orbit was in progress, with a view to adjusting its course towards the comet. Mission control told Keller that any test should be put off until after the encounter. He didn't really expect the camera to work, did he?

When Manfred Grensemann arrived to observe the camera test, on a grey dawn at Darmstadt, he found the science room deserted and only some weary watchkeepers in mission control. Gerhard Schwehm knew that the project manager would be angry, and he had figured out a rearrangement of the links to the spacecraft so that the test could proceed. Keller, who was at his hotel preparing to go home disgruntled, had been alerted.

While Grensemann respected the judgment of the flight operations team on most matters, this was different. Holding his pipe to his chest like a staff of office, he listened to the technical reasons for the proposed postponement and then vetoed it. As long as there was the least chance of obtaining pictures of Comet Grigg-Skjellerup, Grensemann said, no effort should be spared.

The round-trip time to the spacecraft brought a dreamlike, slow-motion quality to all attempts at haste, but the operations team changed the communications band and groomed the spacecraft for the camera test with all the urgency they could muster. Then a command went out to the camera to look straight into the Sun. If it couldn't see that, it must be blind.

As before, the camera's internal systems worked well, but the piece of smashed baffle that had masked its aperture ever since Halley was still in place. There was no reason why it should have fallen off. All that Keller and his team could see of the Sun was a glow of scattered light on one side of the frame, like daylight stealing around a bedroom curtain. After his starring role at Halley, Keller would be just a spectator at Grigg-Skjellerup.

Others were reassured by the outcome. If by a miracle the camera had been operable it would have played merry hell with the power budget at the encounter. To find eleven watts to run the camera would have meant sacrificing other experiments. Fritz Neubauer was content too. This time the jagged spikes of interference from the camera motor would not intrude on his delicate measurements of cosmic magnetism.

CHAPTER ELEVEN

THE GRIGG-SKJELLERUP SYMPHONY

A BIG SHIFT had occurred in the balance of power within the Giotto science team. At Halley's Comet, the camera and the mass spectrometers were the top, Category 1 instruments. All were either wiped out by Halley's dust or made ineffective to varying degrees by the awkward orientation and speed of the spacecraft at the impending encounter. The new mission depended on experiments adopted in Category 2 or 3.

Dust measurements might reveal evolutionary changes in Comet Grigg-Skjellerup, as compared with Halley. Was the proportion of dust to gas higher or lower? Were there any changes in the relative numbers of grains of different sizes? The scientists reviewing the mission in 1990 thought that a practical payoff from observations at Grigg-Skjellerup should be an assessment of the dust hazard for future missions to comets, whether aimed at a long flight in company with a comet, or at a landing to recover samples.

Tony McDonnell's dust-impact detectors were coupled to the bumper shield, which would be oddly angled at the encounter. If Giotto went close enough to the nucleus, he might count a few dozen hits. Anny-Chantal Levasseur-Regourd's Halley Optical Probe Experiment became relatively more important in the absence of other instruments. HOPE could provide no image of the nucleus but, looking out along Giotto's spin axis, it was ready to analyse the comet's light, assess the abundance of dust, and identify selected gases.

Special aspirations lay with the magnetometers of Fritz Neubauer and the particle detectors of Alan Johnstone, Henri Rème and Susan McKenna-Lawlor. The chief motive for Giotto's mission to Comet Grigg-Skjellerup had become plasma physics, which was the formal name for studies of electrified matter and its associated magnetic fields. It was a subject mysterious to laymen, repugnant to news editors and unappealing even to some scientific colleagues.

'Very detailed, very complicated and very dull,' a prominent European space scientist said, making no bones about his opinion of the plasma physics of a comet's interactions with the solar wind, as envisaged for study at Grigg-Skjellerup when Giotto arrived there in July 1992.

'You might say the same about a Bach concerto, if you didn't care for classical music,' another physicist retorted.

Who was right? Was it just a matter of taste, whether you preferred to hang a picture of a comet nucleus on the wall or listen to the melodies of invisible electric particles spiralling in the magnetism of interplanetary space? No, plasma physics was curiously isolated from other sciences, in spite of its contributions to knowledge since the 1920s. That was when the term 'plasma' came into use for a parcel of electrified gas, whether in the tube of a neon sign, the heart of a star or the tail of a comet. Anyone wanting to share the excitement of Giotto's plasma physicists, about the impending encounter with Grigg-Skjellerup, had to let their imaginations cross the cultural divide and enter a realm of wriggly matter.

Plasmas are normal in the Universe at large, and rank with solids, liquids and gases as a state of matter. The Sun's daily fireworks of jets and prominences display the weird ways of plasmas, as do the green drapes and red veils of auroras high in the polar air. But it is only because plasmas are scarce at the Earth's surface that we think them odd.

Although electrically neutral, plasmas seethe with electric currents and magnetic fields. They possess equal numbers of negative electrons and positive ions – electrically charged atoms and molecules – but Hannes Alfvén in Stockholm divined the turbulent love-affair of particles and magnetism in a plasma. Mobile particles stir up magnetic fields that in turn constrain the particles, in a two-way interaction that gives plasmas a liquid-like cohesion denied to ordinary gases.

Magnetism rules when particles from outer space spiral down the lines of the Earth's strong field to make auroras, but the swift particles of the solar wind steal magnetism from the Sun itself and export it into interplanetary space. Any doubt about which is in charge can lead to quarrels. The plasma squirms like an angry snake and the magnetism expresses its discontent in waves. Charged particles bob up and down, or from left to right, as magnetic Alfvén waves pass by.

Misbehaving plasmas chasten the physicists who want to control nuclear fusion by imitating the Sun's fiery core in a magnetic bottle. The writhing stuff finds endless tricks for cooling itself or bursting its magnetic bonds. In advanced experiments, cold pellets of fusion fuel vaporize in a hot plasma, just like comets intruding into the solar wind.

The discovery of the solar wind, predicted from comet tails, made plasma physicists more popular with space managers than with fellow-scientists whose interests lay elsewhere. Particle detectors and magnetometers are cheap and lightweight, and always have something to measure, if not the radiation belts of the Earth or Jupiter then the ever-changing solar weather in interplanetary space.

The plasma of the solar wind, travelling at around 500 kilometres per second, moves so much faster than planets or comets that these objects seem like posts in a fast-flowing river. They part the stream and set up bow waves and wakes. The planet Mercury, airless and non-magnetic, is a triviality for the solar wind to bypass, but the Earth's magnetic shield is ten times wider than the planet itself. It holds the wind at bay in the round magnetosphere on the sunward side, and in the magnetotail stretching downwind of the Earth. In the approach to the magnetosphere the solar wind slows down abruptly at the bow shock, a boundary that curves around the sunward side of the Earth and extends in a great V downwind. Giotto sampled the bow shock, magnetosphere and magnetotail as it swung by the Earth in 1990.

Comets produce their round-headed, long-tailed disturbances in the solar wind by other means. A comet has no magnetism, but it also lacks the gravity to check the outward flight of gases from its atmosphere. When warmed by the Sun, it generates a halo of atoms, molecules and ions which pollutes the solar wind for a million kilometres around. Dirty plasmas also exist in newly forming or exploding stars, but space scientists can study them at first hand in the natural plasma physics lab of the Solar System.

Far from a comet, on the upwind side, the contamination is little more than a burden of mass added to the rarefied but coherent solar wind. As water vapour and other gases escape upwind from the comet, the Sun's rays break them up, liberating electrons and creating ions that are heavier and slower-moving than the typical hydrogen and helium nuclei of the solar wind. As they cross the magnetic force-lines embedded in the solar wind, the charged particles create electric fields that can pick up the ions and accelerate them. The 'pick-up ions' spiral along the magnetic lines and cause friction with the normal wind, which slows down and becomes turbulent. If pick-up ions recover their missing electrons they can shoot back towards the comet as neutral atoms or molecular fragments travelling at high speed.

Close in, the comet's atmosphere is dense enough to exclude the wind and its magnetism, making the 'magnetic cavity' that Giotto found at Halley. With its magnetism severely ruckled around this obstacle, the deflected wind streams past the comet, leaving the cavity tadpole-shaped. Electrified gas from the comet finds its way into the plasma tail by way of a 'contact surface' where the magnetic pressure of the plasma balances the outward pressure of the comet's atmosphere.

When Giotto and the Vegas visited Halley, pick-up and pile-up were plain in the plasma data radioed to the Earth. As the pick-up of cometary ions became a nuisance to the solar wind, quite far from the comet, magnetic waves created turbulence. At the bow shock, the burden seemed to be substantial enough to slow down the solar wind fairly abruptly. And around the magnetic cavity the magnetic force-lines and associated particles were congested in a 'pile-up region'. But these and other features of Halley's collision with the solar wind remained cryptic after six years.

The plasmas gave Nature so many ways to effect the transition from pure solar wind to pure cometary atmosphere, that the scientists could not agree even on the names, never mind the mechanisms, of the various zones of plasma seen at Halley. Inside the bow shock, if that was correctly identified, lay a 'mystery region' half a million kilometres thick, where Henri Rème's electron detector in Giotto recorded prolonged bursts of relatively energetic electrons. Other sensors felt fusillades of energetic ions, amid rapid magnetic fluctuations. Theorists were bewildered.

Near Halley's heart came a trumpet blast of energetic particles which swamped Susan McKenna-Lawlor's detectors at more than a hundred times the normal count-rate. Apart from a component due to charged dust, the only explanation on offer was that ions accelerated in some clash of magnetic field lines in front of the pile-up region, and perhaps also in the comet's tail, might have contributed to the observed fluxes.

The uncertainties were as stark as if a weatherman could not tell whether a high wind was due to a tornado, a hurricane or a forest fire. To be fair, Halley was a gargantuan comet on which to start such studies. The plasma physicists needed another comet, less raucous than Halley and more compact too, so that local gusts in the solar wind itself should not confuse the cometary contributions to the turmoil. Then they might hear the music more clearly.

Optimists thought that Grigg-Skjellerup might be ideal in both respects, although the Sun and the solar wind would be stormier in 1992 than in 1986. The contrast in size promised other bonuses. Grigg-Skjellerup would not be just a small version of Halley to the plasma physicists. Different processes would scale down in different ways, so helping to distinguish the mechanisms at work. Some processes might actually intensify, because the solar wind would collide with the small comet more suddenly and at closer quarters. Magnetic turbulence outside the bow shock could be stronger. There might be new plasma phenomena not seen at Halley.

Giotto's physicists had to keep a rein on their expectations. The comet could be too small and feeble to give good results. In the rate at which gas gushed from the nucleus, Halley outperformed Grigg-Skjellerup by at least a hundred to one. In Giacobini-Zinner, a more active comet than Grigg-Skjellerup, the ICE spacecraft barely detected a bow shock. Grigg-Skjellerup might have none. The magnetic cavity around the nucleus would be less than 200 kilometres wide, compared with 8000 kilometres at Halley. The odds were against Giotto finding its way in.

Never mind. Fritz Neubauer's magnetometers would record the ground-bass of whatever Grigg-Skjellerup might choose to play on encounter day. Alan Johnstone's plasma analyser would pick up the woodwind tones of the commonest ions, warm or cool, and Susan McKenna-Lawlor's EPONA would hear the brass of energetic particles. But the strings might be missing, if Henri Rème's electron detector did not work well after being ripped open by Halley dust.

Giotto and Comet Grigg-Skjellerup converged in the constellation of Leo, near the direction of the bright star Regulus. They rose later than the morning Sun over the eastern horizon of Bavaria, where the 30-metre radio dish of the German space agency's Weilheim station provided mission control's main uplink channel for commanding the spacecraft. As Earth rolled eastwards, Madrid's 70-metre dish could acquire Giotto an hour later. After its sterling service in the reactivations, Madrid would be the main downlink for the spacecraft's signals during the encounter.

When Giotto and its target set in the west, Europe's midsummer evenings were still far too light for seeing the stars, never mind an exceedingly faint comet in Leo. The Calar Alto observatory in Spain, which recovered Grigg-Skjellerup in September 1991, could no longer track it, and neither could an observatory in Arizona which had helped to confirm its orbit. The watch on Grigg-Skjellerup in the last few weeks before the encounter depended on astronomers in the southern hemisphere, who could see the comet after their midwinter sunset. Observatories in New Zealand and Australia, where Messrs. Grigg and Skjellerup came from, played leading parts in the campaign, as did the European Southern Observatory's optical telescopes in Chile.

The astronomers' support was even more necessary than at Halley's interception. There were no Vegas this time, to find Giotto's path into the comet's heart, and aiming for Grigg-Skjellerup depended entirely on the ground-based observations. The nucleus being indistinguishable at long range, its location at the centre of the faint and smudgy comet could only be inferred. The first months of comet-watching left a margin of error of about 1000 kilometres in the computed position at the encounter. Observations in the last few weeks and days almost halved the uncertainty.

The scientists wished Giotto to pass within 2000 kilometres of the nucleus, and much closer if possible. Ideally it would go 200 kilometres downwind of the nucleus, in the tail-forming region where no spacecraft had been before. Knowing that was too much to expect, in view of the estimated errors, the scientists agreed with the flight control team that the best aim-point was the nucleus itself. The odds against hitting it were roughly 100,000 to one.

As Giotto's navigator, Trevor Morley studied two ellipses on his chart. The smaller one, with a maximum radius of 120 kilometres, encompassed all of the spacecraft's possible positions at the time of the encounter, based on the tracking data and its probable errors. The other ring surrounding it, with a radius of 600 kilometres at its widest, showed the possible positions of the comet's nucleus, taking account of the latest observations from the southern hemisphere.

'Aiming for the nucleus' meant nothing more nor less than making the centres of the two rings coincide. The heading was already so good that Giotto's point of arrival had to be shifted by only 145 kilometres to the west. On Wednesday 8 July, two days before the encounter, the radial thrusters took turns to fire as Giotto rotated, thus nudging it sideways into the corrected orbit.

The burn lasted for fifty-two minutes, and changes in the frequency of the spacecraft's signals showed that it fell short of the desired effect by just 2 per cent, or three kilometres from the aim-point. As that was the least of the errors with which he contended, Morley was satisfied with his aim. He slightly tweaked the expected time of closest approach to the comet's nucleus, to some instant between 15:30 and 15:31 zulu on 10 July.

Would anyone be able to observe Grigg-Skjellerup from the Earth at the moment of the encounter? The professionals shook their heads. The only places on dry land from which the comet would then be visible in a dark sky were the islands of Mauritius and La Réunion in the Indian Ocean, which were not celebrated for their astronomical facilities.

On 10 July 1992 the solar wind was very steady as Giotto cruised through it and Grigg-Skjellerup drew near. This augured well for discovering perturbations in the plasma. Had the Sun been stormier, its fluctuations could have masked the comet's effects, just as a rough sea obscures the bow wave of a passing ship.

On the previous evening Antonello Morani had prepared the spacecraft for the encounter, and then switched off the uplink channel to save the precious watts. As at the reactivation in May, the spacecraft operations manager had listed the procedures to adopt if anything went wrong during the encounter. Should the spacecraft go off the air, or if sandblasting by the comet's dust harmed its solar cells, Morani would know just what to do. But some of his contingencies had the forlorn annotation, 'Action: none (no redundancy).'

Giotto was alert, but the first murmur from Grigg-Skjellerup was late and disappointing. The comet seemed even feebler than predicted, as the scientists at Darmstadt watched for data arriving from Giotto on that Friday morning.

At 09.30 zulu, six hours and 300,000 kilometres before the close encounter with the nucleus, Alan Johnstone's plasma analyser registered the long-awaited pick-up ions, telling of contamination of the solar wind by cometary gas. But heavy ions attributable to water vapour from the comet were few and far between. Johnstone estimated that Grigg-Skjellerup was releasing gas at only 60 kilograms per second, or less than one three-hundredth of the emission rate from Halley. He warned his colleagues that disturbances of the plasma, due to the collision with the stuff of the comet, could be muted. They might be observable only within a few thousand kilometres of the comet's nucleus.

As the hours passed, Giotto contradicted him. The counts of pick-up ions in Johnstone's instrument increased faster than could be explained just by the closer proximity of the comet. The initial low counts were from fragments of gas molecules that had travelled outwards from the nucleus for several days. Grigg-Skjellerup was edging nearer to the Sun and becoming more active. The fresher emanations, closer in towards the nucleus, were more generous.

Off an early-morning flight from Munich, a Danish messenger from the headquarters of the European Southern Observatory hurried into the science room at Darmstadt and produced from his bag, in the manner of a conjuror, the latest image of the comet seen from Chile. At 00.05 zulu or 2.05 a.m. Darmstadt time, a German astronomer had inspected Grigg-Skjellerup with the observatory's largest telescope. Despite viewing conditions made awkward by a brightening Moon, a veil of cirrus in the Chilean sky, and the position of the comet low on the horizon, he recorded the very first hint of a dusty tail.

Relayed to Munich in digital form, the image was composed of small squares corresponding with the array of charge-coupled devices used to obtain it. Each

Comet Grigg-Skjellerup as imaged on 10 July 1992, fifteen hours before the Giotto encounter, by the European Southern Observatory's 3.6-metre telescope at La Silla, Chile. The slight elongation of the dust cloud to 30,000 kilometres in a direction away from the Sun (lying top right) was the first hint of a visible tail. Other objects in the image are stars. Observation by Klaus Jockers, Max-Planck-Institut für Aeronomie, Katlenburg-Lindau. (European Southern Observatory.)

square was 630 kilometres across, which meant forty-five seconds of Giotto's flying time through the comet. Overall the image showed the visible head to be 20,000 kilometres wide, but slightly elongated to 30,000 kilometres in a direction away from the Sun. It was a feature never seen before in any picture of Grigg-Skjellerup. The aged comet was greeting the intruder like a galleon spreading its sails for battle.

'Now Giotto has the shout.' With those words, the poet Dante publicized his friend the painter, in *La divina commedia* no less. Some 680 years later, the European Space Agency vowed to make a better fist of telling the world about the encounter with Grigg-Skjellerup than it had managed with Halley's Comet. Preparations for the big day began while Giotto was still in hibernation.

Thunderstorms over Darmstadt supplied a Wagnerian sound-track as senior space officials from Washington and Moscow met their European counterparts from Paris at the European Space Operations Centre, to see how Giotto would fare. Together with reporters and other guests, they were ushered into a large hall, walled with starscapes, that exhibition-makers from Tourex of London had adapted into a television studio.

A full-scale Giotto rotated on a stage, alongside a wire model that represented expected features of the plasma around the comet. In the centre of the studio, the spacecraft's scientific instruments were on display. On a giant video screen and monitors around the hall, the guests could watch pre-recorded material that included animations explaining Giotto's cosmic odyssey. As the encounter proceeded, they also saw real-time interviews with key individuals in mission control and the science room.

The guests found that they were the studio audience for a live TV show. Running for ninety minutes, it included the period of Giotto's closest approach to the comet. Alec Nisbett, formerly a senior science producer for the BBC, orchestrated the programme for the European Space Agency. It was aired directly in the United Kingdom, around the world by the BBC World Service, and in Germany with interventions by a German-speaking reporter. Other stations recorded the TV transmission in order to translate and adapt it for their own feature programmes or news reports.

The presenter was a broadcaster and astronomer, Heather Couper, who burst vivaciously into the studio in a bright yellow suit. She took the stage with the physicist Johannes Geiss from Bern, one of the fathers of Giotto. As the scientific linkman, Geiss explained technical points, interrogated the principal investigators in the science room, and commented on the progress of the mission.

Whatever might befall the spacecraft as the encounter proceeded, the guests, the press and the viewers at home would know it almost as soon as the mission controllers and the scientists. The scraps of often bewildering news, available in the press room at the Halley encounter, were a thing of the past. Whether Giotto itself fizzled or triumphed, the broadcast from Darmstadt would remain Europe's best shot at informing the public in real time about an event in space.

The first remarkable discovery at Comet Grigg-Skjellerup came when Giotto was still half an hour and 25,000 kilometres from the comet's nucleus. Fritz Neubauer's magnetometers were sensing comet-induced waves in the magnetism of the solar wind. They were much stronger and more rhythmic than anything of the kind at Halley. As traces from the magnetic sensors wormed their way down Neubauer's recording paper, the smooth, neatly pointed waves might have been results coming from a lab next door rather than a collision 214 million kilometres away in space.

That such waves could exist, and drink in their energy from charged particles racing outwards from the comet, was known in theory to the plasma physicists, but they had never seen anything like these before. About once a minute, or every thousand kilometres, a wave rose to a crest of strong magnetism. Each was like a long sea-wave, an ocean roller threatening to break as surf on the cometary shore. Every crest had a cap of foam, consisting of energetic electrons bound by the strong magnetism. When Henri Rème remarked that his damaged detector was sensing the electrons in the same waves clearly, the most nervous observers knew that the mission to Grigg-Skjellerup was a success. Even if Giotto suddenly expired, it had found something new in plasma physics.

While all the other sensors sampled the immediate environment of the spacecraft, Anny-Chantal Levasseur-Regourd's HOPE instrument was looking out for the luminosity of a large volume of the comet's head. At 25,000 kilometres, HOPE was already detecting the characteristic glow of charged molecules of carbon monoxide and the hydroxyl fragments of broken water. And, just as the latest image of the comet from the European Southern Observatory had led the investigators to expect, the optical probe began seeing the sunlight scattered by fine dust particles, as the range fell below 20,000 kilometres and Giotto entered Grigg-Skjellerup's visible head.

A chorus came from the particle detectors when Giotto was twenty minutes and 17,000 kilometres from the nucleus. Susan McKenna-Lawlor noted a clear rise in her counts of energetic particles, suggestive of a boundary crossing. Johnstone and Rème were in no doubt that the comet's bow shock washed over their instruments at that time. The bursts of particles were plainer than in Halley's bow shock six years earlier and, as expected, the plasma beyond the shock was markedly slower and cooler than in the unmodified solar wind.

By the time the signals from the bow-shock passage reached the Earth twelve minutes after it happened, Giotto was already 10,000 kilometres nearer its target, in its own time-frame. As at Halley's Comet, events at Grigg-Skjellerup outpaced communications, and for the spacecraft less than ten minutes remained until the closest approach to the nucleus. As in a sea-fight, months of preparation and days of manoeuvre culminated in quickfire action.

As Giotto's sensors hummed with ever-stronger signs of the comet's presence they felt the pile-up of field lines of the solar wind, draping themselves around the exclusion zone of the magnetic cavity. While the magnetism climbed to a level higher than in Halley's pile-up region, the emanations from the comet cooled the plasma. The spacecraft drove recklessly onward, into the most luminous inner region of the comet's head. The unseen nucleus was very near, much closer than at Halley. Would the comet rake the intruder, and blast its unshielded instruments and solar cells with a deadly salvo of dust grains?

The time of greatest danger came as the sky brightened abruptly. Clouds of comet dust illuminated by the Sun bathed Giotto in a glare of light. The purblind spacecraft perceived it with its HOPE instrument, together with a surge in the light telling of the presence of carbon monoxide and hydroxyl. For fifteen seconds, or 200 kilometres along the spacecraft's track, Grigg-Skjellerup looked like a grand, menacing comet. Then the glare faded, almost as fast as it appeared.

As the spacecraft hurtled into this dense region of glowing dust and gas, the silence in one channel was uncanny. Where was the percussion section of the comet's orchestra? The microphones of the dust-impact detection system, on which Halley had beaten an unrelenting tattoo, had heard no hits whatever. The dust seen by HOPE was powdery stuff scattered over a large volume . . . *Accidenti!*

Giotto winced as a dust grain the size of a small pea slammed into it at fourteen kilometres per second. This happened when the nucleus of Grigg-Skjellerup was hurrying northwards past the spacecraft, at the closest approach. With a mass of about thirty milligrams, the impacting grain was comparable with the 'large' grains that shook and injured Giotto at Halley. But with far less energy of motion the shock was more like a bullet than a hand-grenade.

The wobble that ensued was less too – about a tenth of a degree – and the impact increased Giotto's rate of spin slightly. The spacecraft compensated for it by speeding up the despin motor under the radio dish. None of the scientific instruments was damaged, and Giotto calmly sent their readings earthwards as it emerged from the comet on the far side.

News of the impact reached Darmstadt with the usual twelve-minute delay. Tony McDonnell, still waiting for his dust-impact detectors to register a hit, had just remarked that it seemed to be a very clean comet.

'An event!' he cried, pointing to a dot on his computer screen. The others in the science room gave a round of applause. Forty seconds later there was another, smaller hit, and in the end McDonnell's trawl of dust grains totalled three. Even allowing for the state of the sensors, which could now detect only large grains, he had expected a few more than that. With Anny-Chantal Levasseur-Regourd seeing plenty of fine dust, McDonnell's result gave immediate food for thought, about the nature of the dust emissions from an elderly comet.

Confirmation of a shortage of large dust grains came from radio observations of Giotto. Halley's dust reduced the spacecraft's speed noticeably, but Grigg-Skjellerup's caused no braking. The radio tracking showed that the speed faltered by less than a millimetre per second on its way through the comet.

The flight control team was aware of the first large impact, and of the resulting wobble which took care of itself. There was no hiatus in communications this time. None of the possible faults requiring action by the controllers, as listed beforehand by Antonello Morani, occurred during the encounter. They watched Giotto performing like a maestro, and escaping from Grigg-Skjellerup with barely a scratch.

'It seems to be indestructible,' Howard Nye said, in a closing comment in the live television broadcast. 'It's been past two comets, and I think the comets are probably coming off worse than we are with the spacecraft.'

Still patiently charting the comet along its exit path, Giotto reported much the same phenomena as on the way in, although in the reverse order and with a telltale lack of symmetry. The bow shock was farther out, and at this outbound crossing Fritz Neubauer saw its magnetic signature clearly, which he had not done on the way in. This verification was the more heartening because some experts had questioned whether Giotto would detect any bow shock at all. A sequence of fluctuations in Susan McKenna-Lawlor's counts of energetic particles, which had lasted for forty minutes through the comet's head, ended at the same time.

The atmosphere in the science room seemed to some even headier than at the Halley encounter. While several experimenters kept their eyes fixed on their monitors, trying to shut out the din, others chattered excitedly about the results. They were abuzz with the question: 'How close did Giotto go to the nucleus of Comet Grigg-Skjellerup?'

Without the camera, there was no direct way of telling where the spacecraft went. Old-timers recalled that this was the reason for wanting a camera in the first place. A slight chance that Giotto might glimpse the nucleus rushing past, either with its star mapper or with its HOPE instrument, was not obviously fulfilled in the first impression of the data.

So the flyby distance, and its bearing in relation to the Sun, remained matters for deduction. From the flight dynamics room, Trevor Morley expressed his hunch that he had sent the spacecraft within 300 kilometres of the nucleus. Neubauer said that the distance off was more than 100 kilometres, because Giotto skirted the magnetic cavity of the comet's inner atmosphere, where the magnetic field would have dropped, perhaps to zero.

The first of the principal investigators to volunteer a distance was Anny-Chantal Levasseur-Regourd. She had scarcely breathed while HOPE signalled the dramatic brightening and fading of the sky when Giotto was at its closest to the nucleus. It was just what she expected, and she was then able to compare the graph with others prepared in advance for finding the miss distance, based on mathematical models and observations of other comets.

Within minutes of the encounter she estimated that the flyby was at less than 250 kilometres, and put the moment of closest approach within a few seconds of 15.31 zulu. If subsequent analyses confirmed a figure around 200 kilometres, as compared with the previous record for a near encounter with a comet nucleus, 596 kilometres for Giotto at Halley, it was an amazing feat to have accomplished only by ground-based observations and careful navigation.

Levasseur-Regourd was open-minded about whether or not the spacecraft passed on the sunward 'day' side of the comet's nucleus, as it did at Halley. But Neubauer suspected, from the look of his data, that he was seeing Grigg-Skjellerup's 'night' side, away from the Sun and downwind of the nucleus in the solar plasma. While Giotto did not penetrate the 'midnight' tail-forming region, it seemed to have visited what plasma physicists called the comet's 'evening' quarter. In plainer terms, the spacecraft apparently passed on Grigg-Skjellerup's far side, as seen from the Earth, and a little to the east.

'The structure is relatively complicated as compared for example with Comet Halley,' Neubauer said. 'So we may have some other things in the data.'

As with the results from Halley, the main themes noted during the encounter and its immediate aftermath would be filled in with intricate variation and interpretation over the months and years to follow. The scientists would learn every cadence and tremolo in the signalled data, as they replayed the Grigg-Skjellerup symphony again and again in their home laboratories.

Only time would show just what chemical notes might be wrung from the signals sent by Hans Balsiger's ion mass spectrometer. Although sadly out of tune with Giotto's attitude and speed at its second comet, it seemed to have picked up water-related ions, and there were modulations that might harmonize with Neubauer's waves. But there was no doubting the quality of the plasma physics. The clarity and sweetness of a small comet's electromagnetic music, recorded on a calm day in the solar wind, would thenceforward be the norm to which all comets could be compared – with reappraisals pending for the plasma data from Giacobini-Zinner and Halley.

'A textbook encounter!' Gerhard Schwehm exclaimed, as the project scientist for the European Space Agency's latest triumph in deep space.

The thundery rain had given way to a pleasant summer evening. At a table set on the roadway outside the cramped science room, experimenters and controllers sat down together for a celebration supper. They drank German champagne from bottles labelled with the Giotto/Grigg-Skjellerup logo.

'Yes I said, "Switch the bloody thing off!"' David Dale acknowledged, recalling the morning after Giotto's encounter with Halley's Comet. 'Don't forget we'd been living with the project for six years. Senior management was able to take a broader view.'

The project manager of the Halley mission was in Darmstadt for the Grigg-Skjellerup event, in the unaccustomed role of an onlooker at a Giottofest. Away from the TV lights and the high-adrenalin operations rooms, Dale and some of Giotto's other creators had a reunion of their own. John Credland, the terror of the experimenters, was there and so was the French engineer Robert Lainé. Not only did Lainé play a leading part in shaping Giotto, but his insistence on generous supplies of thruster fuel made the extended mission possible.

When Rüdeger Reinhard, project scientist for Halley, saw the new results coming in from instruments he had known from their conception twelve years earlier, he could tell what the wiggles and splodges represented. The others at Dale's reunion were content to share, as no one else could, their deeply personal satisfaction that the bloody thing still worked.

David Link and Rod Jenkins, over from British Aerospace, could take unbounded pride in Giotto. Their comet interceptor, as they liked to call it, executed its original mission to Halley perfectly. It then endured ill-treatment never visualized when it was under construction. The rude business of putting it to sleep and reawakening it twice was sheer improvisation. In its 1986-90 orbit Giotto suffered fierce overheating; in 1990-92 cruel chilling. By Grigg-Skjellerup day it had survived in space ten times longer than the spec required, and was shockingly underpowered. Yet Giotto intercepted its second comet as suavely as if it were brand-new.

The spacecraft was lucky, certainly, but its good fortune was well-earned. On Black Thursday, 22 February 1990, the luck seemed to run out when the wounded Giotto ignored a third command to fire its thrusters. But the engineers cured it and Europe completed the first-ever reactivation of a spacecraft after a long hibernation, in time for the Earth swingby, another 'first'. Leaving aside the damage inflicted by Halley, even the experts were hard put to find anything in the high-tech machine that did not work well. Had the manufacture, assembly and testing by the Bristol team not been meticulous in every detail, a glitch in a vital component such as the despin motor would have cancelled the party at Darmstadt that summer evening.

Dale put the success another way. 'If we had designed Giotto for this day,' he said, 'it would have cost twice as much.'

The flight operations team resumed command of the spacecraft eleven hours after the closest approach, when Giotto had left Grigg-Skjellerup more than half a million kilometres behind it. Between 02.20 and 03.00 zulu on Saturday 11 July, the scientific instruments closed down one by one. The first to go, by a timecoded instruction sent before the encounter, was Anny-Chantal Levasseur-Regourd's HOPE experiment. The last was Fritz Neubauer's magnetometer, extinguished by radio command via the uplink station at Perth in Australia. The scientific mission to Comet Grigg-Skjellerup was over.

Still no one wanted to forsake Giotto. Jokers at Darmstadt had long expressed their hope that it would hit the nucleus of Grigg-Skjellerup and so put a stopper on any talk about a third mission. They remembered with groans the sleepless days and nights after Halley, when they ad-libbed Giotto's return in hibernation to the Earth's vicinity. But early in July the further extension of the mission ceased to be a laughing matter.

The European Space Agency's director of science, Roger Bonnet, told Manfred Grensemann to reprieve Giotto for another seven years. Thanks to the accuracy of the aim towards Grigg-Skjellerup in 1990, the spacecraft used very little hydrazine to fuel its thrusters after the reactivation in May 1992. It still had about fifteen kilograms left, more than enough to put Giotto into an orbit that would bring it back to the Earth's vicinity once more, in 1999.

Again, the orbital correction was a matter of tuning the spacecraft's natural tendency to return close to the Earth at intervals. In this case Giotto should be persuaded to complete eight revolutions of the Sun while the Earth made nine – counting from the swingby in 1990. The control team had the necessary manoeuvres planned before the encounter with the comet. The thruster burn was scheduled for 13 July.

The management team postponed it, to make a final assessment of options and costs. Were the orbital correction to be put off for a few months, it would consume slightly less fuel. At the next Earth flyby very little hydrazine would remain, so saving a few kilograms might make a difference to what could be done with Giotto in 1999. It might still be far too little for any major operation, and the possible gain did not outweigh the problems and costs of keeping the team together and negotiating the necessary ground stations.

On 21 July Darmstadt commanded Giotto, via the Weilheim station, to alter its attitude until it was pointing, tripod first, as nearly as possible along its orbital path. Then the two axial thrusters burned for four hours and consumed ten kilograms of fuel, to fulfil Trevor Morley's calculations for the necessary change in orbital speed. For two days, the ground stations tracked the spacecraft in its new orbit, and then a brief burn adjusted the velocity by three metres per second. When that was done, Antonello Morani released the stream of pre-prepared commands for the third hibernation.

The operation was more complex than before. To make quite sure that the spacecraft was nicely tucked up for its seven-year sleep, mission control remained in two-way contact, via its small antenna, even after it had swung into its spit-roast attitude, with its radio dish turned away from the Earth. At 17.09 zulu on 23 July 1992 their last command went out, and the controllers said *Arrivederci* to Giotto.

The spacecraft would reappear on 1 July 1999 and pass between the Earth and the Moon. If there were no pre-encounter course correction it would fly by the

Earth, 220,000 kilometres out, at 02.40 zulu and approach within 160,000 kilometres of the Moon at 13.20.

Seven years left plenty of time for considering what to do with Giotto then. One idea prevailing when it went to sleep for a third time was to treat it as a technological relic. If the spacecraft were partially reactivated at close range, engineers planning future missions would be fascinated to diagnose the state of its various components after fourteen years in space, from the solar cells and batteries to the charge-coupled devices of Uwe Keller's camera. Switching on some of the instruments for a scientific flyby of the Earth, as accomplished in 1990, was another possibility. The Moon, too, was an aim-point to think about.

The near-exhaustion of the thruster fuel, reduced to an uncertain few kilograms, seemed to rule out anything more ambitious, including the gossip about 'a third comet' heard on every hand at the Grigg-Skjellerup encounter. But Giotto had confounded the pessimists so often that nay-saying seemed as rash as over-optimism. Who knew what ingenious plan the flight dynamicists might come up with by the end of the decade? And who could tell what bonuses Nature might put in the path of an indestructible comet raider?

'So will you be back here when Giotto gets to its next target?' Susan McKenna-Lawlor was asked at Darmstadt.

'Do you think I'd be anywhere else?' the Irish experimenter retorted.

The US space shuttle Atlantis put the Europeans' Eureca-1 into orbit a couple of weeks after Giotto closed down, so ESOC was far from idle. The same shuttle crew wrestled for the first time with an Italian tethered satellite, the brainchild of the late Giuseppe Colombo of Padua, who was influential in the creation and naming of the comet mission. Exactly 500 years before, in the summer of 1492, another Colombo, Cristoforo, was on the launch pad beside the Atlantic in Spain, counting down to his voyage to the west.

The gifts from Europe's explorers and innovators in the intervening centuries often seemed incurably poisoned by cruelty and strife. Indeed, this tale of Europe's first venture into the Solar System started in the aftermath of the Second World War. If anyone still doubts whether people of many nations, whose parents were bitter enemies, can work together in harmony, tell them about Giotto.

The engineers and scientists are decades ahead of the politicians. The spacecraft's components, remember, came from ten countries and pieces of its scientific instruments from almost as many. Europe's own Ariane rocket, masterminded in France, sent Giotto on its way. Of the project managers and project scientists for the two phases of the dual mission, three out of the four happened to be German. A British spacecraft operations manager flew Giotto into Halley's Comet and a young Italian took it through Comet Grigg-Skjellerup. The management of such a polyglot enterprise would startle many businessmen, yet it was entirely routine for the European Space Agency, as for other major scientific collaborations in Europe.

And if anyone supposes that multinational action is bound to be arthritic and bureaucratic, let them recall how Europe conceived Giotto at great speed, in a race against Halley's Comet and in defiance of American shillyshallying and scepticism. The élan was still there in 1986, in the snap decision to keep the battered spacecraft alive after Halley for a second mission. As David Dale likes to say, 'Giotto is people.' They represent Europe's new renaissance more truly than all the conclaves of paper-pushers put together.

First and last, it is a scientific mission. Towering over the technical and managerial accomplishments is the house that Giotto built: a treasury of new knowledge. Its splendour makes sense of all the sleeplessness and stress endured by a thousand Europeans amid the runways of Bristol and Toulouse, the tulip fields of Noordwijk, the rain forest of Kourou and the mainframes of Darmstadt – and in the scattered laboratories that now rival the cathedrals as embodiments of the old continent's loftiest spirit.

'This achievement places the European Space Agency at the forefront of cometary exploration,' Roger Bonnet declared, about Giotto's dual mission. The intelligence from Giotto and its erstwhile companions, concerning the character and contents of comets, is now so thought-provoking that scientists long to hold pieces of a comet in their hands before they die.

CHAPTER TWELVE

CHILDREN OF THE COMETS

In previous centuries, comets seemed the most interesting objects in the sky because they were big and mobile. Now they fascinate scientists because their nuclei are so small. They preserve, scarcely modified by heat or pressure, the ingredients from which they were assembled billions of years ago. These wandering time-machines are more complex in composition than any inanimate material ever studied before.

Some future spacecraft may shadow a comet nucleus in its extended orbit, as it sleeps in cold regions far from the Sun and then, when it feels the solar warmth, rouses itself to spew gas and dust into a head and tail far larger than itself. But a greater prize will come when a spacecraft lands on a comet nucleus, gathers samples of its ice and less volatile materials, and flies home with them to the Earth.

A sample-return mission represents the highest objective for comet science at the present state of knowledge. Its supporters reason that comet-stuff will illuminate our origins more dazzlingly than anything else within the reach of spacecraft. Even without surprises, the discoveries from a sample-return mission to a comet cannot fail to be exciting. The tantalizing message from the Giotto mission and other recent studies is that comets probably hold the key to our very existence. In some sense still to be defined precisely, we are children of the comets.

More prosaically, one can say that thorough laboratory analyses of cometary material should provide firmer information on several salient subjects:

1. The origin of the chemical elements.
2. The origin of the Solar System.
3. The origin of the Earth's ocean and atmosphere.
4. The origin of carbon compounds on the Earth.
5. The origin of life.
6. The origin and character of Earth-threatening objects.

A comet sample-return mission is therefore one of the most important ventures now on offer, not just in cometary space science but in the whole of fundamental research. The European Space Agency has for several years planned to despatch a spacecraft called Rosetta to fetch comet samples, early in the 21st Century. Rosetta is now in jeopardy, but that contentious issue is deferred to a postscript. The question is not whether such a mission will be attempted, but only when and by whom.

The present chapter provides a commentary on comet science after Giotto, in telling how comet samples will help with the subjects listed, and therefore in understanding humanity's kinship with comets and with the Universe at large. A remark about the composition of our bodies sets an agenda for explanation.

In our chemistry at least, we are beasts from outer space. Along with the rest of the Earth's cargo of life, we use elements that are very common in the Universe but relatively scarce in our rocky home. Lists of the most numerous atoms look like this:

EARTH'S CRUST	SUN	HALLEY'S COMET	HUMAN BODY
Oxygen	Hydrogen	Hydrogen	Hydrogen
Silicon	Helium	Oxygen	Oxygen
Aluminium	Oxygen	Carbon	Carbon
Magnesium	Carbon	Silicon	Nitrogen
Iron	Nitrogen	Nitrogen	Calcium

The conventional opinion is that life was lucky to find its necessary elements on the Earth, even in limited amounts. A more assertive view is that contributions from space painted the planet with the ingredients of life.

The origin of the chemical elements (from the Big Bang to about 4.6 billion years ago). That we are made of stardust is a proposition not queried since the 1950s, when astronomers and nuclear physicists figured out how stars of different sizes and conditions manufacture the various elements of which we and our surroundings are built. What's new is the ability of scientists to particularize the story in respect of our own elements, by identifying ancestral stars. The stellar archaeologists have done wonders with meteorites, and comet samples returned to the Earth will give them a far richer source of material.

The Big Bang itself made our hydrogen atoms, together with some helium, but virtually all of the other elements were forged in the nuclear furnaces of stars. Sunlight, and most starlight, comes from thermonuclear reactions that consume hydrogen and make more helium. When an ageing star runs out of hydrogen at its core, the helium burns, making carbon and oxygen. These burn in their turn to form neon, magnesium, silicon, sulphur and iron. Carbon-rich and oxygen-rich stars inject grains of carbon, silicate and silicon carbide into interstellar space.

That is how Nature assembles its commonest elements. To make elements heavier than iron-nickel calls for more drastic action. Giant stars lead short but brilliant lives and then explode as supernovas. They scatter bountiful supplies of the common elements into space, but they also drench the materials in subatomic particles. These build atoms all through the periodic table of the elements, to uranium and beyond.

The products of this alchemy taint the primeval hydrogen gas with grains of icy, carbonaceous and stony stardust, and with chemical vapours detectable by radio telescopes. Dark dust clouds obscure the view of distant stars in the Milky Way. Thanks to generations of stars that grew old and exploded before the Solar System came into being, the mother cloud that gave birth to the Sun was doped with elements capable of making rocky planets like the Earth, and living things.

Gravity caused the mother cloud to collapse and squeezed its materials into the Sun and the planets. As if in a giant blender, the traces of the ancestral stars lost their identity. A lump of iron, for example, probably contains atoms from dozens of anonymous stars.

Such was the state of play until scientists studied peculiar white grains in a meteorite that fell in Mexico in 1969. Many of the lumps of stone or metal that drop out of the sky as meteorites come from collisions among the asteroids swarming between Mars and Jupiter. These are rocky bodies, and heat and pressure have changed their materials from their pristine state. But other meteorites, rich in carbon, seem to have a different, more gentle history. Some may be pieces of former comets.

The Mexican meteorite was carbon-rich, and the white grains turned out to be older than the meteorite that carried them – older than the Sun and the Earth. The atomic composition showed the material to be made of dust from a star that blew up shortly before the formation of the Solar System. The shock of the explosion may have triggered the collapse of the mother cloud.

In this case the surviving grains are up to a centimetre wide. A second breakthrough in stellar archaeology occurred in 1987, when a team led by Edward Anders of the University of Chicago discovered, again in meteoritic material, ancient grains visible only in a microscope. Besides specks of graphite and silicon carbide, there are also very small diamonds. As Anders puts it, 'They'd barely make engagement rings for bacteria.'

An instrument called an ion microprobe finds in each grain the chemical hallmark of an individual star. No two stars are exactly alike in their proportions of elements and of isotopes – atoms of the same element with different masses. The variations distinguish the grains from the chemical averages of the Solar System, and from one another. Atomic impurities created by cosmic rays during the billions of years of each grain's existence give an impression of its age. Hundreds of different stars may be distinguishable in due course.

A comet is a safer storage place than asteroids and meteorites, for the pre-solar grains. Anders has therefore been a strong supporter of the European Space Agency's proposals for the Rosetta sample-return mission to a comet. His colleague Ernst Zinner, of Washington University in Missouri, who led the ion microprobe investigations of the meteoritic grains, has also contributed to Rosetta's scientific planning. The mission could advance stellar archaeology far beyond its present pioneering stage.

Discovering where the atoms of the Solar System came from is only part of the payoff expected from stardust in a comet. Astronomers will make better sense of their distant observations of exploding stars, interstellar dust and star-forming clouds, when they can see examples of the materials. Distinctive stellar grains will give astrophysicists intimate examples of the nuclear processes by which stars build the elements. Investigators will also look for rare grains going back 10 billion years or more, to when the Universe and the Milky Way Galaxy were very young and their chemistry was much more primitive.

The origin of the Solar System (from 4.6 to 4.5 billion years ago). Given a stock of elements scattered in space, the next question is how they were assembled into the Sun and its family of planets, including the Earth. The importance of comet samples in this respect is that they may represent the only well-preserved mementos of the cloud of dust and gas called the solar nebula, which surrounded the newborn Sun and gave birth to the planets. Chemical and physical changes in the Earth and other planets have so altered their ingredients that scientists can only guess, or at best compute, their original forms.

New stars are still forming in the Milky Way Galaxy where we live, at a rate of perhaps one a year, and all of the stages can be seen in various parts of the sky. Astronomers observe dense patches forming in molecular clouds. They see complex processes of collapse accompanied by fierce outward-acting jets, and stars coming alight for the first time, just as the Sun did about 4.6 billion years ago. Infra-red telescopes find disks of dust surrounding some of the stars.

In the solar nebula that swirled around the newborn Sun, the dust included stony and icy solids inherited from interstellar space. Theorists picture dust gathering into boulders, and the boulders into microplanets, or 'planetisimals'. In a welter of increasingly massive collisions, the microplanets merged to make planets. Where they failed to do so, in the Asteroid Belt between Mars and Jupiter, microplanets remain as relics of the planet-building process. Analogues of the process are found in the systems of moons and rings that surround the giant planets. The Earth's solitary and disproportionately big Moon may have formed from debris hurled into space by the impact of a large object on the newly-formed Earth.

Ice could not survive in the hot inner regions of the solar nebula. There the solids were stony, with high contents of silicon, oxygen and iron. Water survived only in the form of hydrated crystals. From these refractory solids originated the small rocky planets: Mercury, Venus, Earth, Mars and most of the asteroids.

Beyond the Asteroid Belt lay the edge of a cosmic Antarctica, where ice abounded in the solar nebula. Ice may have been the main material for making the cores of Jupiter, Saturn, Uranus and Neptune. These were massive enough, especially in Jupiter and Saturn, to attract and bind large quantities of hydrogen and helium gas from the solar nebula.

Comets are microplanetary relics from that outer region, where they swarmed in unimaginable numbers at the birth of the Solar System. A million million comets would weigh no more than Uranus and Neptune. To purge the skies of the left-over comets required swingbys and collisions with the large, outlying planets. While many comets were ejected right out of the young Solar System, sufficient found themselves in the far-flung orbits of the Oort cloud to keep the inner Solar System re-supplied with comets for billions of years.

How can samples of a comet, returned to the Earth for analysis, help to prove or vary this story? The composition of the samples will give scientists an unprecedented snapshot of the solar nebula in the region where the comet formed. Data on gases trapped in the ice will give an impression of the distance from the Sun. For Halley's Comet there is already an approximate answer.

As Giotto's mass spectrometers confirmed, carbon monoxide is second only to water vapour among the most abundant materials in the comet's atmosphere. Some is emitted from dust grains, but the rest of the carbon monoxide is evidently released from the comet's water-ice as it vaporizes in the Sun's heat. Ice forming from water molecules at very low temperatures (below 137 K, or minus 136 degrees Celsius) makes an irregular matrix called 'amorphous ice' which is well able to capture other molecules and retain them.

Akiva Bar-Nun and Idit Kleinfeld of Tel Aviv have tested the trapping of gases in ice at various low temperatures. From the abundance of carbon monoxide in Halley's ice they conclude that the ice amassed at about 50 K, or minus 223 degrees Celsius. The temperature suggests that Halley's birthplace was in the Uranus-Neptune region of the solar nebula, at twenty to thirty times the Earth-Sun distance. If Halley is built of blocks that came together gently, as Uwe Keller concludes from the Giotto images, the blocks must have been on very similar orbits at that range from the Sun.

Giotto's high-speed impressions of Halley's atmosphere have much wider margins of error or ambiguity than will occur when laboratory analyses of comet samples become possible. The very process of emission of gas and dust from the comet, whereby they reached the spacecraft, altered the materials from the condition in which they had lain undisturbed inside the comet for billions of years. Here, as in other respects, Giotto and the other comet missions have whetted the scientists' appetite, rather than satisfying it. It is a typical state of affairs, in the most productive science.

As comets are creatures of the outer Solar System, any connection with the origin of the Earth might seem indirect at best. But the swingbys of the outer planets sent many of the surplus comets plunging sunwards. The inner Solar System was ablaze with comet tails, in a veritable Age of Comets that lasted several hundred million years.

Consequences of the inevitable collisions with the newborn Earth will emerge in the next section. A remark about the young Sun sets the scene. Astronomers see newly formed stars, in the so-called 'T-Tauri phase', emitting extremely strong winds. When the Sun passed through that phase, it must have swept the Solar System with a solar hurricane. Comet samples may reveal individual particles from the Sun, captured in the material at the time.

The T-Tauri wind was strong enough to strip away any primitive liquids and gases that existed on the surfaces of the Earth and the other small, rocky planets at the time of their formation. The present ocean and atmosphere therefore must have originated at a later stage, after the Sun settled down and became better behaved. The question now to be addressed is where that second, more durable stock of liquids and gases came from.

The origin of the Earth's ocean and atmosphere (from about 4.5 to 4 billion years ago). When a space package carrying comet-stuff eventually parachutes home, it may be splashing down into an ocean of melted comets. The beer consumed to celebrate the event in the recovery ships may be mainly comet-stuff too, even though it will owe its special qualities to recent activities of yeast. Among some space researchers an opinion is hardening, that the water in the ocean, and the small quantities doled out to the atmosphere, ice sheets, rivers, lakes and so on, came mainly from the ice of comets that rained down on the young Earth.

Life runs on the loose materials at the Earth's surface. Plants grow by using sunlight to combine water with carbon dioxide, while bacteria supply them with nitrogen extracted from the air. Animals take in oxygen to release energy from their food, but they require water even more urgently than the food. Where the water and gases came from is therefore a matter of intimate concern.

For more than thirty years, intermittent suggestions have come from scientists, that the Earth and other small planets and moons acquired veneers of gases, vapours, fluids and ices from impacting bodies. The notion seemed literally far-fetched at first, but Giotto gave new life to the proposition and a sample-return mission may help to clinch it.

Space scientists thinking this way defy the geologists, who have been taught to believe that the ocean and atmosphere formed from volcanic steam and gases gushing from the Earth's interior. A breakdown of hydrated crystals in deep-lying rocks supposedly released the water. Volcanic emanations do include steam, but nowadays it is nearly all recycled surface water that percolated into the crust and boiled out again. To account for the large volume of water in the ocean, and to say how an ocean and atmosphere existed when the Earth was still young, geologists have to imagine a great burp of water vapour and other gases from the rocks of the interior, at an early stage of the planet's history.

The idea that comets were at least an important contributor of water is no more conjectural than that. Nor does the evidence from Halley give any reason to reject it. At least one can now say that the comet's water is of the same kind as the Earth's.

Our water carries a genetic fingerprint in the heavy hydrogen, or deuterium, that it contains. For every 12,700 atoms of ordinary hydrogen, there is one atom of deuterium. Small though it is, this proportion of deuterium in the hydrogen of the Earth's water is more than ten times higher than in the commonplace hydrogen gas of the Universe. It is also more than five times higher than in

Jupiter or Saturn, which drew most of their hydrogen directly from the gas of the solar nebula.

Results from Giotto's neutral mass spectrometer, interpreted to figure out the proportion of deuterium in Halley, suggest a degree of enrichment similar to that in the Earth's ocean. Peter Eberhardt of Bern and his colleagues say cautiously of this Giotto finding, that it 'does not contradict' theories that comets supplied a veneer of materials to the Earth and its planetary neighbours.

Molecules containing deuterium are more sluggish in their behaviour than those with ordinary hydrogen, and that is what makes enrichment possible. The best place for it to happen is in the dark spaces between the stars, where very cold molecular clouds are relatively far richer in deuterium than the Earth. The solids of the solar nebula may trace back to cold interstellar dust grains and ice particles.

In the opinion of Tobias Owen of Honolulu, a noted investigator of planetary atmospheres, the solids created a reservoir of deuterium-rich materials distinct from the deuterium-poor gas that came into the solar nebula. The consequences are plainest in the contrast between the gassy planet Saturn and its icy moon Titan, which is remarkable for its thick atmosphere. Measurements for the same compound of hydrogen (methane) reveal an almost tenfold increase in deuterium from Saturn to Titan. This makes sense if Saturn sucked in most of its hydrogen from the gas of the solar nebula, while Titan acquired it from the solids.

In this perspective, the similarity in deuterium between the Earth's ocean and Halley does not settle the question of origins. Hydrated minerals, akin to those built into the Earth as an internal source of water, turn up in meteorites. They show the same enrichment in deuterium, typical of the solids of the Solar System. But indirect evidence for the contribution of comets to the ocean comes when Owen traces the origin of the Earth's atmosphere.

Life interacts comprehensively with the main gases of the air, and the best guide to the origin may be the traces of the chemically inert 'noble' gases in the atmosphere. Of these, the helium, argon and radon come mainly from the decay of the Earth's stock of radioactive elements. The neon, krypton and xenon need a more cosmic or geochemical explanation, as does a lightweight form of argon, argon-36. These noble gases are much scarcer on the Earth than in the Sun and their proportions are different, too. Atoms of argon-36 outnumber atoms of xenon-132 by nearly 100,000 to one on the Sun, but on the Earth by only 1000 to one. This looks like another genetic fingerprint. The mystery deepens with the discovery that Venus, the Earth's sister planet, has a greater proportion of argon-36, making it more like the Sun in this respect.

Owen explains the patterns by deliveries from comets. He draws support from laboratory tests of the trapping of gases in ice, by the Tel Aviv scientists mentioned earlier, Akiva Bar-Nun and Idit Kleinfeld. Suppose you wanted to use comet-ice to smuggle on to the young Earth the krypton and xenon and depleted argon-36, in exactly the proportions they have in the atmosphere today. What should the temperature be in the solar nebula, where the ice amasses? The experimental answer is spine-tingling: 50 K or minus 223 degrees Celsius, just the temperature at which Halley's Comet formed, according to the data for carbon monoxide.

Such genetic fingerprints make a sensitive cosmic thermometer. A shift of only a few degrees up or down from 50 K would leave the proportion of argon trapped in the ice noticeably wrong, compared with the Earth's inventory of noble gases. This prompts Owen to offer an explanation of the high argon-36 in the atmosphere of the planet Venus.

A giant comet that formed farther out in the solar nebula, at or below 30 K (minus 243 degrees Celsius), could trap enough extra argon to top up Venus's quota in a single impact. The necessary diameter of the comet nucleus would be about 120 kilometres. That is far larger than Halley, yet smaller than Comet Chiron which orbits far from the Sun and is about 200 kilometres wide.

Argon-36 comprises 44 parts per million (by weight) of the Earth's atmosphere, or 22 billion tonnes. Delivery by comet-ice would mean the deposition of larger quantities of water on the Earth at the same time. But how much? Could it be 100 million times the argon-36, as required to match the 1.4 billion billion tonnes of water present at the Earth's surface? Only direct measurements in cometary material would answer the question. Even then, much would depend on the condition of the ice in the samples, because comet-ice sheds many of the trapped gas molecules if its temperature rises above 137 K, or minus 136 degrees Celsius.

The Age of Comets provides a timeframe when deliveries of comet-water to the Earth could have taken place. A prime discovery from lunar exploration is that an intense bombardment of asteroids and comets assailed the Moon for hundreds of millions of years after its formation. Most of the large pockmarks and blotches on the lunar surface turn out to correspond with a series of violent impacts in the period 4.4 to 3.8 billion years ago.

The Earth suffered a bombardment far more severe than the Moon's. With its stronger gravity, it captured many more of the passing asteroids and comets and caused them to hit harder. Geology healed the scars of that early era, but the water, gases and other soft material at our planet's surface may commemorate the events. The lunar evidence shows the rate dwindling, with perhaps 95 per cent of the impacts occurring before 4.0 billion years ago. That is the best estimate for when life began, by which time the Earth must have been well watered.

Putting rough numbers to the bombardment, the mass of material hitting the young Earth in the form of large impactors may have been about 10 billion billion tonnes. If half of the mass consisted of comets, and half of each comet consisted of water-ice, and half of the material was blasted back into space by the force of the impacts, then the delivery of comet-water would have been about 1 billion billion tonnes. Extremely rough though it is, that figure invites comparison with the 1.4 billion billion tonnes of water at the Earth's surface.

'Being the right size is as important as being at the right distance from the Sun,' Owen says, 'if you want to be a planet that can harbour life.' On small planets and moons gravity is too weak to prevent gases and vapours escaping into space. In the past, astronomers considered only the heat from the Sun and the planet's interior, as the driving force that could expel the gases. In the more violent circumstances now envisaged, the kinetic energy of impacts could also throw the gases into space. A very heavy or fast impact by a comet or asteroid could strip a planet of gases painfully acquired in many lesser impacts of comets.

Mars is a case in point. Its atmosphere is far thinner than the Earth's and Owen blames most of the loss on 'impact erosion'. This possibility confuses an already tortuous story about atmospheric origins. Even if the solid Earth did burp gases into the early atmosphere, subsequent heavy impacts could have removed them.

Skirting this and many other complications, the case for a cometary origin of the ocean and atmosphere has been sketched here as if it were an all-or-nothing issue. To deny that any water or gases of the Earth boiled out of its interior would be as stupid as to say that comets delivered nothing. The goal is a new history of the ocean and atmosphere that will assess the changeable supplies from comets and from outgassing, throughout geological time. Only then will scientists be able to say just what fraction of the beer-water came from somewhere near Uranus.

A sample-return mission to a comet is a prerequisite. Atom-by-atom counts of the noble gases and the isotopes of hydrogen, carbon, nitrogen and oxygen will make the fingerprinting more precise. Titan, the cloudy moon of Saturn, is another key target for testing the theory of the cometary veneers, and the European Space Agency is planning to land a package of instruments on Titan.

The verdict is therefore postponed for some decades, on how much of the Earth's ocean and atmosphere was contributed by comets. Confident space scientists like Owen predict that it will be 'Most'.

The origin of carbon compounds on the Earth (from about 4.5 to 4 billion years ago). As water makes up 59 per cent of the mass of the human body, confirmation that most of it came from cometary impacts would go a long way towards establishing us as children of the comets. But life is pre-eminently a chemical trick of carbon atoms. They have the ability to join with one another, and with hydrogen, oxygen, nitrogen and other elements, to make molecules of virtually unlimited complexity and subtlety.

Life began in a primeval soup, according to the general opinion of those who speculate about the event. The soup was water enriched with fairly elaborate carbon compounds emanating from 'prebiotic' chemical reactions. As visualized by Alexander Oparin in the Soviet Union back in the 1920s, the first organisms supposedly assembled themselves from the materials of the soup. They then fed on the soup until such time as they found other ways of sustaining life.

The question of where those carbon compounds came from is worth separating from the origin of the atmosphere (previous section) and the transition to life itself (next section). Regardless of whether comets were involved in those episodes, a distinctive case can be made for a cometary role in making the soup.

For many years, scientists assumed that the Sun's ultraviolet rays, and the action of lightning strokes, provoked chemical reactions of the necessary kinds. In a famous experiment in Chicago some forty years ago, an electric spark passed through a mixture of simple gases and created a brown tar. The compounds it contained included amino-acids, the sub-units of protein molecules. But the starting mixture was peculiar. Except for water vapour, the gases used were quite different from those common on the Earth today. They were hydrogen-rich materials: hydrogen itself, methane (hydrogenated carbon) and ammonia (hydrogenated nitrogen).

That mixture now looks unbelievable, as a simulation of the Earth's early atmosphere. The carbon was more probably in the form of carbon dioxide, and the nitrogen as nitrogen gas. The production of elaborate compounds of carbon by ultraviolet rays and lightning would have been far less efficient than in the Chicago mixture. Deliveries of carbon compounds from space would then have provided most of the flavouring of the primeval soup.

Giotto and the Vegas confirmed that Halley is rich in carbon compounds, but the impacts of comets on the young Earth would be a poor way of delivering them. Delicate carbon compounds in the cargo would succumb to the intense heat released in an impact. The accompanying shock in the atmosphere would, though, stimulate the formation of other carbon compounds from the ingredients of the atmosphere itself, thereby supplementing those produced by lightning strokes and ultraviolet rays.

The biggest extraterrestrial source of carbon compounds may have been very small interplanetary dust grains swept up by the Earth. Many of the grains would have come from comet tails, but they would descend more gently through the atmosphere. Christopher Chyba and Carl Sagan of Cornell University estimate that far more material reached the Earth's surface in the form of dust than by direct cometary impacts. They also calculate that extraterrestrial carbon compounds predominated over 'home cooking' in the primeval soup, if less than 10 per cent of the early atmosphere took the form of hydrogen gas.

The most direct guide to the nature of the incoming carbon compounds is the interplanetary dust grains themselves. These can be harvested in small numbers from the stratosphere and, on average, carbon compounds make up about a tenth of their mass. More generous examples of the carbon compounds likely to have found their way on to the grains, and thence into the primeval soup, will become available when samples from a comet nucleus are returned to the Earth.

The origin of life (about 4 billion years ago). What will scientists be keenest to check, when they have pieces of a comet in their hands at the end of a sample-return mission? Surely those discoveries which suggest that we may owe our very life to a comet. Of all the ideas that have come to scientists while digesting the results from the Halley space fleet, the boldest is that an impacting comet may have triggered the origin of life, by injecting substances into the primeval soup.

This hypothesis goes well beyond the notion that comets and comet dust made important contributions to the young Earth's stock of loose materials and carbon compounds. It nicely balances, without contradicting it, the proposition that comets and comet-derived asteroids interrupted the subsequent evolution of life with their deadly impacts. And it makes cometary chemistry a matter of more than astronomical curiosity.

One of its advocates is Jochen Kissel, whose instruments analysed Halley's dust. The big crew-cut physicist at the Max-Planck-Institut für Kernphysik at Heidelberg was always something of an outsider in the Giotto science working team. This was partly because Kissel had instruments in the Soviet Vegas to think about too, and partly a matter of temperament.

An earlier chapter told how Kissel and his chemist-colleague Franz Krueger took the results from the dust mass spectrometers in Vega-1 and Giotto and inferred a wide range of carbon compounds. On the basis of their discoveries, they formulated their idea of how life began. Their hypothesis plunges like its comet into a sea of opinions and prejudices about the origin of life.

All the wonders of the living planet come from a conspiracy of string-like carbon compounds: nucleic acids, which carry information, and proteins which, as enzymes, do most of the work in keeping organisms alive. The genetic code used in the conspiracy is shared by every creature, whether a bacterium, a rose or a whale. This suggests that all are descended from the same simple ancestral cells which arose in the primeval soup. But even at its simplest, life is an intricate chemical stunt and easily snarled.

In the 1970s Manfred Eigen, a German chemist, theorized that the earliest ancestor of all life was a 'hypercycle' of cooperating molecules. He supposed that a set of nucleic acids and proteins came together by chance in the soup and then multiplied at the expense of other molecules. Laboratory tests showed a form of nucleic acid appearing spontaneously and out-reproducing all its rivals. At first, this test-tube evolution relied on the presence of a protein, an enzyme that aided the replication of nucleic acids. Later, the enzyme turned out to be unnecessary. Zinc could do the trick.

Eigen's nucleic acids are not the famous DNA of human heredity, deoxyribonucleic acid, but its poor relation RNA, or ribonucleic acid. The oldest types of molecules in our bodies are made of RNA and they may be 'living fossils' from the ancestor of all life on Earth. They perform workaday tasks in decoding genetic messages and manufacturing proteins. Sub-units of RNA are often required to assist the enzymes that catalyse vital chemical reactions. A boost to Eigen's idea that RNA was the key to life's origin came in the early 1980s, when American molecular biologists found that RNA itself could catalyse certain chemical reactions. This gave it enzyme-like powers previously assigned only to proteins.

Even so, other experts continue to assert that proteins, not nucleic acids, were the harbingers of life. Some theories invoke mineral materials as catalysts for early life, or even as progenitors in the form of self-reproducing crystals. But chicken-or-egg arguments about which came first, nucleic acids or proteins, are not the main problem for those who want to argue that life began unaided by 'home cooking' on the young Earth.

Irrespective of the precise sequence of events, an improbable collection of active chemicals had to assemble itself by chance. Even if random reactions went on for hundreds of millions of years in every drop of water in the world, awkward chemical laws reared up like monsters in the soup to prohibit the crucial step to life. The active molecules had to be much more concentrated than their food in the soup, yet if an aggregation occurred by chance it could persist only for a moment before thermodynamic laws of disorder abolished it.

Moreover, RNA and proteins required concentrated energy for their assembly, yet the soup had to avoid high temperatures if it were not to destroy the delicate chemical strings. Existing life brings cool energy to bear by trapping sunshine or by feeding on energy-rich molecules of previously living foodstuffs. To concentrate the necessary energy in a primitive cell on the young Earth seems as inconceivable as running a car on a lukewarm consommé. Krueger and Kissel's hypothesis offers, from a comet, the biochemical equivalent of gasoline.

Krueger and Kissel are not the first to call in extraterrestrial help for bringing the world to life. In 1907 the Swedish chemist Svante Arrhenius imagined free-range bacterial spores driven through interstellar space by the pressure of starlight, like the dust of comet tails. Eventually, he said, some spores alighted on the Earth, and seeded it.

Arrhenius dodged the question of how and where the bacteria arose in the first place. So did Francis Crick, co-discoverer of the structure of DNA, who in 1971 joined with Leslie Orgel in reasoning that intelligent beings in another part of the Galaxy might have deliberately seeded the Earth with bacteria sent in spaceships. So did Fred Hoyle and Chandra Wickramasinghe, when in 1981 they visualized bacteria escaping naturally from one planet, and travelling to others aboard comets. Krueger and Kissel, on the other hand, see life originating on the Earth with assistance from space.

As the Halley probes showed, the contents of a comet are chemically hungry, and ready to react vigorously when brought into contact with water and other warm materials. Krueger and Kissel visualize a comet delivering its celestial carbon compounds suddenly and in small packets into the ocean of the young Earth. The contrast between the bland soup and the comet-stuff provides chemical energy for assembling new materials, exactly where it is needed.

Picture a comet rushing into the atmosphere of the early Earth. There is an explosion and most of it vaporizes. Fragments splash into the deep ocean, where they slow down gradually. They include fluffy mineral grains just a few thousandths of a millimetre wide and coated with the reactive carbon compounds identified in Halley. A piece of a comet no bigger than a football, surviving the impact and entering the ocean, contains a million billion such grains.

In principle, a single grain with the right chemistry is enough to start life on Earth. The surfaces of its mineral components, rich in metal ions, can promote chemical reactions. Zinc, also seen in Halley, helps in the assembly of RNA strands. The fluffy and porous mineral grain provides a ready-made compartment in which large, active molecules can remain locked away in high concentrations that contrast with the dilute soup. The pores admit small molecules from the soup to nourish the reactions. These include amino-acids suitable for making proteins.

The magic collection of materials then has to make its own compartments, so as to be anchored for ever to the mineral grain. The electrochemical contrast with the surrounding sea-water encourages molecules of fat to wrap themselves around the concentrated chemicals, making a primitive cell wall. Like the mineral grain, the fatty envelope protects the RNA while letting small molecules in and out.

The cell-like blob grows and soon splits into new blobs, each containing copies of the key RNA strings. Provided it is more successful than anything produced from other grains, its offspring can take over the entire ocean in a matter of months. If one comet fails to do the trick, perhaps because its impact is too violent, there will be another along soon, in the Age of Comets when life begins. All the rest of the story of life is then just a matter of time and evolution.

Life starts quickly and easily, according to this hypothesis. It should therefore occur on virtually any planet with liquid water that is exposed to cometary impacts. Although the Earth is the only planet in the Solar System that fits the bill (at least nowadays) there would be every reason to expect some planets of other stars in our Galaxy to possess living creatures. Whether they would include intelligent, communicative species, of the kind that radio astronomers keep listening for, is quite another question.

In the Krueger-Kissel hypothesis, if the Earth was the mother of life, a comet was the father. Neither could fashion living cells without the other. The hypothesis would make us the children of the comets in a more genetic sense than the propositions that we are made of comet-stuff or that cometary impacts made our evolution possible.

Life's origin is far too important to treat lightly, and many other theories, fads and fancies have come and gone. Right or wrong, this version of how life may have started is a powerful spur to sending a spacecraft shopping in the Solar System for samples of a comet. If cometary sperm may have impregnated Mother Earth, hadn't we better look for them under the microscope?

The origin of Earth-threatening objects (from 4 billion years ago and into the future). After the origin of life, the Age of Comets ended. The rain of comets and asteroids on the Earth became far less intensive, but it never ceased entirely, and will continue in the future unless human beings prevent it. To be children of the comets assumes yet another meaning, in the evolution of life on the Earth.

We humans are a highly improbable product of random and aimless lurches in evolution. With total indifference, Nature created and then wiped out trilobites, tyrannosaurs, toxodonts, and millions of other species. Cosmic impacts caused mass extinctions of animals, and many abrupt changes in the course of evolution seen in the fossil record. If any of a long succession of comets or asteroids responsible for major events had passed the Earth safely, twenty minutes earlier or later, the chances are we would not be here. Evolution would have followed other courses. And a big impact could bring the human experiment to an abrupt end.

Ever since Edmond Halley himself suggested, 300 years ago, that Noah's Flood was caused by an impacting comet, it has been obvious to astronomers that objects blundering through the planetary traffic of the Solar System present a threat to living things on the Earth. The discovery in 1932 of Apollo, the first dark asteroid to be seen orbiting near the Earth, introduced another kind of hazard.

Unlike the Moon, the Earth buries its impact craters, so they are much harder to find. More than a hundred are known, though, ranging up to 200 kilometres in diameter. Only in a handful of cases is there a clear one-to-one connection between a mass extinction and a large impact crater of the same date. But astronomers looking for geological evidence of cosmic impacts found them in the fossils, and in the sudden changes in life that punctuated the geological periods. They said so even before the discovery of a world-wide dusting of exotic material just at the time when the dinosaurs died out.

Biologists and fossil-hunters read different books from the astronomers and were slow to learn. Those who refuse to accept that cosmic impacts mattered for evolution keep faith with their predecessors who denied that the continents moved. Some palaeontologists accept the impact scenario and wonder whether disappearances of species not associated with global catastrophes might be due to local effects of smaller impacts. Another question is whether there is any rhythm in the pattern of extinctions through geological time.

When Jan Oort proposed a distant cloud of comets as the source of 'new' objects, close encounters with passing stars were seen as the chief means of dislodging swarms of comets from the cloud and into the inner Solar System. Other destabilizing agents now range from massive clouds to tides in the Milky Way Galaxy.

Some of those who chronicle the history of life from the fossil record assert that major extinctions have occurred every 26 million years. In the mid-1980s there was speculation about a Nemesis Star, on a distant orbit around the Sun, that perturbed the Oort cloud every 26 million years and sent in a shower of comets. Searches have failed to reveal the star, and most astronomers doubt if such a distant companion could remain attached to the Sun. Those who still believe in the 26-million-year cycle want the astronomers to find another explanation.

Comet experts see little need for special showers to account for the major extinctions. The near-Earth asteroids Betulia and Ivar are eight kilometres wide. Halley's Comet is bigger and hurtles the wrong way around the Sun. These are current examples of objects that could easily cause a stupendous global disaster like that which extinguished the dinosaurs. Be reassured that neither they nor any other object so far identified will collide with the Earth during the next few centuries, or as far ahead as the orbits have been computed.

A reckoning of the present threat from outer space, based on the present populations of comets and near-Earth asteroids, comes from a NASA workshop. An estimated 2000 asteroids are in potentially dangerous orbits, although fewer than 200 have been identified so far. The small number of short-period comets like Grigg-Skjellerup, passing in small orbits near the Earth, increases the impact hazard due to these asteroids by only 1 per cent.

Comets arriving unpredicted on large orbits are another matter. The NASA workshop's definition of a major impact, capable of causing a global catastrophe, corresponds with an asteroid more than 1 kilometre wide hitting the Earth with an impact energy greater than 100,000 megatons of TNT. Such events are estimated to occur at intervals of about 100,000 years. About three-quarters of them involve near-Earth asteroids, but the rest are due to long-period comets that come 'out of the blue' and travel at higher speeds than the asteroids.

A sample-return mission to a comet will make valuable comments on the impact hazard, past and future, starting with the origin of the near-Earth asteroids. Some have wandered in from the Asteroid Belt beyond Mars, but are others dead comets, as many experts believe? A hands-on assessment of cometary material will help scientists to tell whether or not it can remain essentially intact after a comet loses its ice in repeated visits to the Sun. Uwe Keller's interpretation of Giotto's image of the nucleus of Halley's Comet implies that it can.

The physical strength of a comet's material will also give an impression of its likely behaviour during an impact. A comet (or a dead comet) may break up more easily than a rocky asteroid during an encounter with the Earth, and so produce a number of simultaneous impacts. Examples of multiple craters of the same age occur in the Earth's crust.

Small fragments of a comet could cause powerful air-bursts. The mysterious explosion in 1908 that flattened forests in the Tunguska region of Siberia, without producing a crater, is now judged to have been the work of a piece of a comet less than a hundred metres wide. It exploded in an air-burst high in the atmosphere and released energy equivalent to a twelve-megaton H-bomb. A comparable impact over a large city could flatten it, set it on fire, and kill a million people.

Events of that size, and equivalent impacts by rocky fragments like the one that created the relatively small Meteor Crater in Arizona, one kilometre wide, will occur far more frequently than planet-devastating impacts, perhaps every few centuries. To confirm that the interpretation of the Tunguska explosion fits with the character of cometary material is therefore another task. And when scientists know the nature of a comet's nucleus better, after a sample-return mission, they will be able to help in making plans for comet-busting operations.

It was after many years of pressure that concerned scientists finally caught the attention of politicians. The NASA workshop already mentioned was part of the space agency's response to a demand from the US Congress for concrete suggestions about protecting the Earth against comets and asteroids. It proposed the creation of a network of large, asteroid-hunting telescopes called Spaceguard – a name taken from a story by Arthur C. Clarke.

A threatening asteroid will be easier to deal with than the most dangerous comets, provided it is identified in good time. Its orbit is predictable for many years ahead, well in advance of any likely collision with the Earth. There should be plenty of time to plan and accomplish a diversion. For example a mass-driver rocket, attached to the asteroid and powered by solar energy, could shoot rocks in a chosen direction for years on end, and so alter the orbit. The rocks would be quarried from the asteroid itself. If the object is a dead comet, it may be easier to rupture into large pieces on divergent orbits. This consideration gives a possibility of practical application to abstruse-seeming studies of the decline of Grigg-Skjellerup and its partners.

Much faster action will be demanded if the Spaceguard telescopes see a long-period comet suddenly appearing on a collision course with the Earth. They are unlikely to spot it more than a few months before the impact, and an explosion may be the only hope for deflecting or smashing the comet, either by a collision with a massive intercepting spacecraft, or by nuclear warheads. If space engineers are to prepare a rapid-response system for such an eventuality, they will want every scrap of data from Giotto and all subsequent comet missions, to help them compute the effects of their missiles on the fabric of the comet nucleus.

POSTSCRIPT

RUNNING HOME WITH THE ICE CREAM

After Giotto's encounter with Grigg-Skjellerup, comet scientists who ask 'What next?' found themselves in the same fix as in the winter of 1979-80. Giotto was born out of a cancelled mission, amid arguments about scientific priorities and budgets, and uncertainty about the roles of the space agencies of the US, the Soviet Union and Europe in exploring Halley's Comet. A dozen years later, and at the time of writing, a similar muddle surrounds the next big comet mission. This book ends as it began, in the curious blend of science, technology, finance and international relations that comprises space politics.

Early in 1992, NASA cancelled a mission called CRAF. Until then, scientists who had gambled decades of their lives on the prospects for space missions to comets thought they had won the jackpot. They had half-promises of two major missions. Besides CRAF, which was intended to make a true rendezvous with a comet and fly in company with it for thirty months, the Rosetta spacecraft of the European Space Agency was to land on another comet and bring samples back to the Earth for analysis. Now CRAF is dead and Rosetta is on the danger-list.

Halley's Comet is not accessible now, to hurry people along with its inexorable schedule. But plenty of short-period comets, cousins to Grigg-Skjellerup, are always in the offing. As a NASA mission with German and Italian participation, CRAF was to have gone to Comet Tempel 2 – a target of bad omen, it seems, because that was also the goal of the International Comet Mission cancelled in 1980.

Indeed, CRAF had much in common with that predecessor. The name was an acronym for Comet Rendezvous and Asteroid Flyby. The asteroid Mandeville was to replace the long-departed Halley as the flyby target, but the main idea was still to shadow Tempel 2 for a long period, closely observing its surface and emissions as these changed when the comet neared the Sun. The solar-electric propulsion visualized for the International Comet Mission was still lacking, but two swingbys of Venus and one of the Earth were supposed to help CRAF find its way into the same orbit as Tempel 2. It was expected to reach the comet in December 2002, after a journey of nearly seven years, and then operate in its vicinity until after Tempel 2's closest approach to the Sun, early in 2005.

The comet-sample return project, Rosetta, is designated as a cornerstone mission in the European Space Agency's long-term science programme. It has a launch date set for 19 April 2002, according to a baseline study published in 1991. The postulated target is Comet Schwassmann-Wachmann 2, although Comet Hartley 2 is another candidate. Swingbys of Venus (once) and the Earth (twice) are supposed to boost Rosetta into an orbit to intercept the target far from the Sun. After a few months of approach manoeuvres and inspections of the comet's nucleus, the spacecraft settles gently on its surface.

The smart electronics of the landing vehicle recognizes the site selected for its touchdown. Rosetta anchors itself to the comet and collects ten kilograms of samples of icy and non-icy materials, by drilling into the surface. The material goes into a refrigerated container designed to keep them below minus 140 degrees Celsius – cold enough to preserve any gases trapped in amorphous ice. From then on, the Rosetta mission is a matter of running home with the ice cream before it melts.

Rosetta takes off from the comet, leaving part of its structure behind, and arrives home on 31 May 2011. An 'aerocapsule' including the refrigerated container separates from the main spacecraft. In a final flight of two hours, the capsule plunges into the Earth's atmosphere and parachutes into the Pacific Ocean. Shipborne helicopters recover the capsule and its hard-won cargo of comet stuff.

The real science begins then. Laboratory equipment, too large, heavy and delicate to fit in a spacecraft, scrutinizes the comet samples grain by grain, molecule by molecule, atom by atom. Such analyses of Moon rocks, delivered to the Earth by the US Apollo astronauts and the unmanned Soviet Luna 16 spacecraft, revealed far more about the Moon's nature and history than any visual inspections or remote experiments could do. Microprobe techniques for differentiating and analysing very small grains have advanced impressively since then. Only by such examination in laboratories on the Earth can the full meaning of a comet's chemistry be decoded.

Rosetta has as good a name as Giotto's. It comes from the multilingual stone which was the key to deciphering Egyptian hieroglyphics. Europe's space scientists hope that comet samples will be similarly revealing about our cosmic origins.

Will Rosetta happen? The European Space Agency's allocation of 550 million accounting units (at 1992 prices) is only about half of what is needed for the mission. The assumption has always been that NASA would supply the launching rocket and the main spacecraft, and control the mission via the Deep Space Network. Europe's share would be Rosetta's lander facility, its complex drilling and sampling equipment, and the sample-return capsule.

Gerhard Schwehm, the project scientist of the Giotto Extended Mission, is also ESTEC's study scientist for Rosetta. His attention was divided in 1992, between seeing Giotto safely to Grigg-Skjellerup and discussing its successor. Angelo Atzei is study manager at Noordwijk, and the Paris office of Roger Bonnet, the agency's director of science, is deeply involved in reassessments of Rosetta.

American attitudes are enigmatic after the demise of CRAF. Common sense might say it clears the way for Rosetta, by reducing the US commitment from two major comet missions to one. But having cancelled a mission initiated by American scientists, NASA and the federal government will not automatically support one promoted by the Europeans.

Political, economic and technical questions complicate the picture. Will the end of the Cold War make space exploration harder or easier to finance? How will the long-lasting competition for funds between manned and unmanned missions develop in the 1990s? What is the role for the Russians, expert but poor, in international space ventures? Or for the Japanese, or Chinese? How will the recession of the 1990s, aggravated in Europe by the cost of re-assimilating Germany's eastern territories, affect enthusiasm for scientific space projects? Will a new American preference for small, cheap and fast missions discourage ambitious long-term ventures like Rosetta? Or will the power of ground-based telescopes now under construction, which will obtain interplanetary pictures as good as the Voyagers', tilt the choice of missions towards landers and sample returns?

The uncertainties leave Rosetta like a bride waiting for the American groom. For fear of a jilting, a study of an all-European alternative of using the dowry for a different mission, was conducted early in 1992. It assumed a launch in 2002 or later, using an Ariane 5 rocket.

One option investigated was an abbreviated CRAF-type mission: a rendezvous with Comet Schwassmann-Wachmann 3 far from the Sun, followed later by two months of observation. Most of the new scenarios involved a rendezvous with a near-Earth asteroid, with the option of finally setting down instruments on its surface. In the most ambitious case considered, the spacecraft would scoop a sample of material from the asteroid and bring it back to the Earth.

Asteroids are not comets, although some may be the corpses of comets. They are not uninteresting, least of all the near-Earth asteroids that present a risk of collision. The trouble is that there are perhaps ten kinds of asteroids, so a visit to one or two would be inconclusive. Compared with a comet, an asteroid is hard to investigate chemically, because its emissions are small or non-existent. And even a sample return from an asteroid might bring back to the Earth nothing more remarkable than the materials to be found in an ordinary meteorite.

European experts on the Solar System have given a thumbs-down to any such 'descoped' Rosetta. The scientific argument for the comet sample-return mission was won, they feel, when Rosetta was adopted as a European cornerstone.

At the time of Giotto's invention in 1980, many scientists advising the European Space Agency took the view that planetary missions were too expensive and should be left to the Americans. Uwe Keller recalls that those supporting Giotto were careful to say, 'A comet is different, so don't be afraid – we're not opening the door to planetary missions.' But Giotto's success did open it. The agency has committed itself to a joint venture with NASA, called Cassini/Huygens, to send a spacecraft to Saturn and put a European lander on the large moon Titan. It has studied possible contributions to the exploration of Mars. But Rosetta is meant to be the jewel in the crown of Europe's interplanetary programme.

The size of the comet-science community has increased, especially in the US. Ground-based and satellite observations of Halley, often funded by NASA, lured many academics into the field. 'Now when you go to a meeting,' Keller observes, 'there are 500 people, where in the 1970s there were fifty.'

The time still seems ripe for a major comet mission to build on Giotto's successes. Supporters of Rosetta are encouraged by news that NASA's Jet Propulsion Laboratory is studying a comet lander for a possible launch in 2002 or thereabouts. It seems a natural partner for the European sample-return scheme. If agency chiefs can handle the diplomacy, and American and European comet scientists avoid the sparring that took place in 1980, Rosetta may be saved.

For all the reasons given in the last chapter, comet samples remain the greatest scientific prize imaginable, for any interplanetary venture of our time. Sooner or later, a sample-return mission of the Rosetta type will fly, as a necessary successor to Giotto. Then wonders may appear, not seen since the world began. Take, for example, the idea about the origin of life inspired by the Halley discoveries. In a lab somewhere in Europe, perhaps in the year 2011, an experimenter may drop a grain of comet dust into a jar of lukewarm soup and watch it come to life.

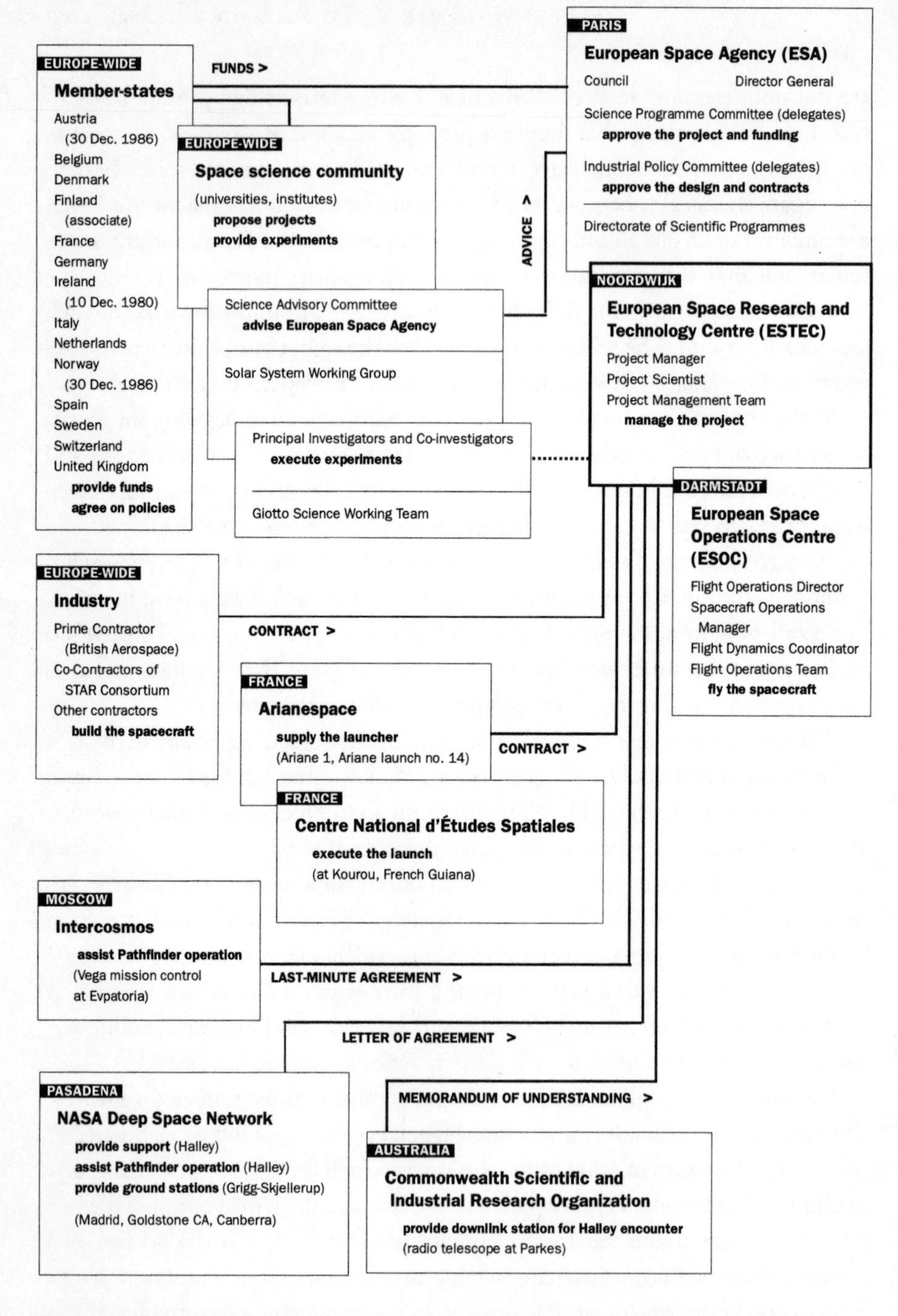

The organization of the Giotto mission. All lines converge on the project management at ESTEC. *Contractual arrangements of various kinds were made with manufacturers and with other space agencies. The relationship with the scientists (dotted line) was close but not contractual.*

INDEX